AIR POLLUTION AND CLIMATE CHANGE

THE BASICS

This book identifies four key forms of air pollution: indoor, urban, regional and global. It discusses how these four types of pollution are manifest in today's society and examines the scientific and policy challenges that stand in the way of progress.

Writing in a style that balances scientific underpinnings with accessible language, Pearson and Derwent examine the sources and historical context of air pollutants, before dedicating a chapter to each of the key forms. Armed with these basics, they begin to address the challenges faced by improving indoor, urban and regional air quality, while reducing global warming in the years ahead. This leads to a greater understanding of the challenges of global climate change, with new proposals for reducing global warming. However, the authors conclude that it is only when we have a scenario of reforestation combined with reductions in emissions of all greenhouse gases that real progress will be made in the fight against climate change. Then, air pollution will also be consigned to history.

With a foreword written by Professor James Lovelock, this book will be of great interest to students and scholars of climate change and environmental policy, as well as air quality professionals working in this important field.

John K. Pearson enjoyed a 30-year career in research with Royal Dutch Shell, including two European assignments. Since his early retirement from Shell, John has been an air quality consultant to industry, working on the development of emissions inventories and exploring air quality opportunities, as well as encouraging research councils to promote user applied science. He is the author of the book *Improving Air Quality*, published in 2001, which builds upon his experience in both Europe and the United States on energy companies, motor manufacturers and legislators working together. His most recent scientific contribution in 2019 was 'Improving Solvent VOC Emissions Inventories', published in *Atmospheric Environment.*

Richard (Dick) G. Derwent took an MA degree and a PhD in physical chemistry from Queens' College in the University of Cambridge. He has spent much of his scientific career studying air pollution and atmospheric chemistry. Dick is one of the foremost experts within the fields of air pollution, atmospheric chemistry and global climate, and has published over 550 peer-reviewed scientific papers, reports and book chapters. He is the joint author of two books: *Atmospheric Pollution and Environment Change* and *Mechanisms of Atmospheric Oxidation of the Alkanes.* He is a visiting professor in the School of Geography, Earth and Environmental Sciences at the University of Birmingham.

The Basics

The Basics is a highly successful series of accessible guidebooks which provide an overview of the fundamental principles of a subject area in a jargon-free and undaunting format.

Intended for students approaching a subject for the first time, the books both introduce the essentials of a subject and provide an ideal springboard for further study. With over 50 titles spanning subjects from artificial intelligence (AI) to women's studies, *The Basics* are an ideal starting point for students seeking to understand a subject area.

Each text comes with recommendations for further study and gradually introduces the complexities and nuances within a subject.

REAL ESTATE
JAN WILCOX AND JANE FORSYTH

MANAGEMENT (SECOND EDITION)
MORGEN WITZEL

SEMIOTICS (FOURTH EDITION)
DANIEL CHANDLER

CHOREOGRAPHY
JENNY ROCHE AND STEPHANIE BURRIDGE

LANGUAGE ACQUISITION
PAUL IBBOTSON

AIR POLLUTION AND CLIMATE CHANGE
JOHN PEARSON AND RICHARD DERWENT

INFANCY
MARC H. BORNSTEIN AND MARTHA E. ARTERBERRY

For a full list of titles in this series, please visit www.routledge.com/The-Basics/book-series/B

AIR POLLUTION AND CLIMATE CHANGE

THE BASICS

John K. Pearson and Richard G. Derwent

earthscan
from Routledge

Routledge
Taylor & Francis Group
LONDON AND NEW YORK

Cover image: © Getty Images

First published 2022
by Routledge
4 Park Square, Milton Park, Abingdon, Oxon OX14 4RN

and by Routledge
605 Third Avenue, New York, NY 10158

Routledge is an imprint of the Taylor & Francis Group, an informa business

British Library Cataloguing-in-Publication Data
A catalogue record for this book is available from the British Library

Library of Congress Cataloging-in-Publication Data
A catalog record has been requested for this book

ISBN: 978-1-032-27517-8 (hbk)
ISBN: 978-1-032-27518-5 (pbk)
ISBN: 978-1-003-29313-2 (ebk)

DOI: 10.4324/9781003293132

Typeset in Bembo
by Taylor & Francis Books

Access the Support Material: http://www.routledge.com/9781032275185

CONTENTS

FIGURES

TABLES

FOREWORD

James Lovelock CH CBE FRS

Physics is often regarded as the queen of the sciences, and it was not surprising to find the prediction of something as important as the Earth's future climate treated as a problem in physical meteorology. In reality we now think that the climate of the Earth is dependent not only on atmospheric physics but also on human and biological changes in the Earth's surface and atmosphere. As discovered by Edward Lorenz, even purely physical attempts to explain climate by mathematical models can give chaotic solutions, which are useless for prediction. Lotka suggested that prediction would be easier if the biological part of the environment was included along with the physical. This conjecture was confirmed in the 1960s, when fast computers became available, by Lovelock and Watson.

This book by Pearson and Derwent clearly shows how much more complex is the real Earth. It lists various factors such as the physics and chemistry of the oceans, and similarly the growing influence on the climate, of people in their cities and by their agriculture, and how in turn this changes the climate. Also, how these changes should be measured.

It is a book for politicians and government servants who really want to know what affects the climate and how they might ameliorate adverse change modestly and without ruinous expenditure.

PREFACE

We have devoted a significant part of our scientific careers to understand the challenges posed by air quality and global warming. John, for his part, has worked in industry to inform his colleagues about these issues and to formulate strategies that will lead to air quality improvement. Dick, for his part, has worked on the basic science underpinning urban, regional and global air quality. Both of us fervently believe that it is not possible to understand air quality and global climate issues fully, unless and until you can put numbers into some basic mathematical equations that describe them. Despite the complexity, there are some hidden basic truths and principles that are worthy of description. These are the basis of our monograph, where we seek to quantify the link between air quality and global warming.

In writing our monograph, we have drawn material from some key basic reference sources, which are listed at the end of each chapter. In addition, we have taken material from a large number of literature papers and web sources. We have checked all these sources as best we can, but we take full responsibility for any errors and mistakes that still remain. Over the years, we have learned much from our scientific colleagues and are grateful for their inspiration. We hope this monograph, in some small way, repays our debt of gratitude to our colleagues.

ACKNOWLEDGEMENTS

After long careers in industry and government, we decided to spend most of our time during the Covid lockdown researching and writing this monograph on the link between air quality and global warming. There are many documents on air pollution and papers on global warming, yet few had examined the link in detail or explored both by means of mathematical calculations. Quantifying this link has been our challenge, with the objective of demonstrating that global warming can be reversed given time.

Over many years, we have worked within many teams of experts from the energy and chemical industries, from international organisations and from individual governments. It has been our privilege to share our knowledge with, and to learn from, many gifted scientists, industrialists, and environmental experts.

We have particularly appreciated detailed technical discussions with Nadine Allemand of CITEPA, Paris, Zig Klimont of IIASA, Vienna, Luc Turkenburg of the European Chemical Industry, Alan Curtis of AERE Harwell, Michael Jenkin of Atmospheric Chemical Services, Okehampton, Harvey Jeffries of the University of North Carolina, Chapel Hill, Tom Wigley of the University of East Anglia and Michael Prather of the University of California Irvine.

Our thanks go to Alan Wilson, who has recommended improvements to our monograph as we have progressed.

We record our deep appreciation to James Lovelock CH CBE FRS for writing the Foreword to our monograph.

It takes enthusiasm and perseverance to write a book and we express our gratitude to our own families for their encouragement and interest.

Air pollution and global warming encompass many fields of science, as well as economics and politics. It has been our privilege to explore how our planet will overcome the fundamental challenges of improving air quality and reducing global warming.

ABOUT THE AUTHORS

JOHN K. PEARSON

John K. Pearson was born in St Andrews in Scotland. He returned to read applied mathematics at the university, and was awarded a United College Bursary in Science, graduating in 1969 as well as winning a class medal in applied mathematics.

He enjoyed a 30-year career in research with Royal Dutch Shell including two European assignments. He was project leader for gasoline specification development in the mid-1980s when he published and presented papers to the Society of Automotive Engineers in the United States and Canada. During this period, he had been responsible for liaison with European vehicle engine manufacturers on fuel developments as well as being chairman of the Fuels Committee of the British Technical Council.

Bilingual in French and English, John was group leader for fuel and lubricant research and development for Shell in France in the early 1990s, responsible for a large international team.

The final five years of his Shell career were his most challenging as principal scientist responsible for air quality research. In this role, he was directly involved in the European Auto Oil Programme and worked closely with legislators and European member states on the EC Ozone Directive in the late 1990s.

Since his early retirement from Shell, John has been an air quality consultant to industry, working on the development of emissions inventories and exploring air quality opportunities, as well as encouraging research councils to promote user applied science. He has presented his work worldwide and persuaded Industry to be pro-active in improving air quality and the environment.

He is the author of the book *Improving Air Quality* published by the SAE in 2001 which builds upon his experience in both Europe and the United States on energy companies, motor manufacturers and legislators working together. His most recent scientific contribution in 2019 was 'Improving Solvent VOC Emissions Inventories', published in *Atmospheric Environment.*

Air quality and global warming encompass everything from science to politics and economics. Careful negotiation between all interested players is critical – indeed having been chairman of both his local parish council and his local hospice, John Pearson's negotiating skills and team playing have been kept well to the fore.

RICHARD DERWENT OBE

Richard (Dick) Derwent took a MA degree and a PhD in physical chemistry from Queens' College in the University of Cambridge. He has spent much of his scientific career studying air pollution and atmospheric chemistry.

He joined the scientific civil service in 1971, initially in the Air Pollution Division of Warren Spring Laboratory. There he set up the first automatic air pollution monitoring network in the UK, measuring NO_x, CO, SO_2, VOCs, O_3 and lead particles. He discovered photochemical smog in London and published the first national emission inventories of NO_x and VOCs.

In 1974, Dick joined the Environmental Medical Sciences Division of the Atomic Energy Research Establishment, Harwell. There he built the first European models of stratospheric ozone layer depletion, photochemical ozone formation, acid rain and the build-up of greenhouse gases: methane and ozone. He was appointed by the UK Department of the Environment as Chairman of the UK Photochemical Oxidants Review Group. He provided the leadership for the scientific underpinning of the negotiations within the UN ECE of the NO_x, VOC and Gothenburg Protocols to the Convention on Long-Range Trans-boundary Air Pollution (CLRTAP) in Geneva.

In 1990, Dick joined the Department of the Environment as the head of the Air Quality Science Unit. There he oversaw the establishment of the UK urban air quality monitoring network providing real-time air quality data to the general public. He was a Lead Author on the first report of the Intergovernmental Panel on Climate Change and developed the concept of the Global Warming Potential (GWP) to express the equivalence between the different greenhouse gases. He constructed the basket of greenhouse gases for the UN Framework Convention of Climate Change (UN FCCC), and this is still in use to this day.

In 1993, he moved to the Atmospheric Processes Research Branch of the UK Meteorological Office as an individual merit scientist. There he was responsible for building the first coupled ocean-atmosphere-atmospheric chemistry climate model. In 1994, he was awarded the Royal Society of Chemistry Award for Environmental Chemistry for his major contribution to the understanding of the processes leading to the formation of photochemical oxidants. In 1996, he was awarded the Fitzroy Prize of the Royal Meteorological Society in recognition of his becoming one of the foremost experts on atmospheric pollution within Europe. He was appointed Officer of the Order of the British Empire (OBE) by Queen Elizabeth II for his work on air pollution and atmospheric chemistry.

In 2003, Dick took early retirement to become an independent scientist studying air pollution, atmospheric chemistry and climate change. He was appointed chairman of the EMEP Task Force on Measurement and Modelling, completing Europe-wide assessments of regional air quality for the CLRTAP. In 2007, his contribution to the award of the Nobel Peace Prize to the Intergovernmental Panel on Climate Change through his work on GWP concept was officially recognised.

Dick is one of the foremost experts within the fields of air pollution, atmospheric chemistry and global climate and has published over 550 peer-reviewed scientific papers, reports and book chapters. He is joint author of two books: *Atmospheric Pollution and Environment Change* (Hodder Arnold) and *Mechanisms of Atmospheric Oxidation of the Alkanes* (Oxford University Press). He is a visiting professor in the School of Geography, Earth and Environmental Sciences of the University of Birmingham.

1

AIR POLLUTANTS AND THEIR SOURCES

OVERVIEW

Air pollutants may be present in the atmosphere as gases or particles. Most air pollutants are directly emitted into the atmosphere, but some are formed there by chemical reactions. Air pollutants are transported across regions and countries and may cross national boundaries. In this chapter, we discuss air pollutants, their health impacts and their principal sources. We also highlight the main greenhouse gases and pollutants: carbon dioxide, methane and ozone. While there is much focus on the reduction of the emissions of those particles which can penetrate the lungs, reducing aerosol particles also affects global warming.

Coal and oil combustion are major causes of air pollution problems, followed by road transport using gasoline and diesel fuels. Massive reductions in carbon dioxide and all pollutants will be achieved over the next two decades by transforming power stations and road transport. It is, however, important to consider all sources of pollution, both anthropogenic and natural biogenic, in order to improve air quality.

Earth's atmosphere contains 78.08% nitrogen (N_2), 20.95% oxygen (O_2) and 0.93% argon (Ar). The remaining atmospheric constituents are called trace gases and include water vapour (H_2O), carbon dioxide (CO_2), neon (Ne), helium (He), methane (CH_4), krypton (Kr) and nitrous oxide (N_2O). Water vapour concentrations are highly variable and average about 1% at the Earth's surface.

DOI: 10.4324/9781003293132-1

The atmosphere is essential to life on Earth, providing the oxygen we breathe, carbon dioxide for plant growth through photosynthesis, ozone to absorb damaging ultraviolet radiation from the sun and a greenhouse blanket of water vapour and carbon dioxide to maintain the habitable temperatures required to sustain life on Earth.

In addition to the major gases and the trace constituents mentioned above, the atmosphere contains a whole range of atmospheric pollutants which are present in trace amounts and whose levels are highly variable in space and time.

The terms air pollution and air pollutant have precise meanings. An air pollutant is a substance which when emitted into the atmosphere causes damage to human health or the wider environment. Air pollutants may be present in the atmosphere as gases or particles. Examples of gaseous air pollutants include sulphur dioxide, nitrogen oxides, carbon monoxide, carbon dioxide, methane and other organic compounds. Examples of particulate air pollutants include smoke, smuts, fine particles from traffic and coarse particles such as sea salts and wind-blown dusts.

Here we examine the common air pollutants and describe their sources in the air that we all breathe.

CARBON MONOXIDE (CO)

This compound is formed by incomplete combustion of fuels linked to insufficient oxygen. Outdoors, the main sources are gasoline vehicles and forest fires. In practice, modern gasoline vehicles emit very small amounts of carbon monoxide and this source has become a problem of the past in air quality. Bonfires and forest fires do give rise to high levels of carbon monoxide, but these are usually brought under control quickly. Indoors, carbon monoxide remains a significant problem. These issues are covered in Chapter 3.

CARBON DIOXIDE (CO_2)

Atmospheric carbon dioxide concentrations are increasing steadily due to combustion processes linked to oil, gas, coal and wood. This is the most important greenhouse gas associated with global warming and climate change and its control is essential in the years to come. Power stations in Europe are phasing out the use of coal and oil in favour of renewable technologies and motor manufacturers are introducing hybrid and electric vehicles. Both these measures are helping to reduce emissions of this pollutant. However, Asian countries continue to build coal-fired power stations and car ownership is increasing rapidly based on gasoline vehicles, generating huge emissions of carbon dioxide. The challenges to reduce global carbon dioxide emissions are covered in Chapter 6.

SULPHUR DIOXIDE (SO_2)

Sulphur dioxide, an acidic gas, is associated with the burning of fossil fuels to generate heat and electricity. SO_2 is readily transported across international boundaries by wind fields which contributes to the challenge of reducing acid rain and its damaging impacts on soils and surface waters. In general, it is in decline, but countries which rely on power stations burning coal such as China, India, Czech Republic, Poland and Serbia still have substantial SO_2 emissions. Low sulphur fuels have also helped Europe to reduce SO_2 pollution over the past 30 years.

NITROGEN OXIDES (NO_X)

Nitric oxide (NO) and nitrogen dioxide (NO_2) are always found together in the atmosphere and classed as Nitrogen Oxides (NO_x = NO + NO_2). In developed countries, the largest source of NO_x comes from the automotive sector, accounting for about 50%, with power stations contributing about 20%. In previous years, the diesel engine has made the greatest NO_x contribution in the automotive sector, but advances in diesel engine technology (for example: de-NO_x catalysts) have started to reduce this considerably. Further reductions continue as the diesel engine gives way to electric and hybrid engine technologies. Most pollutant sources emit their NO_x as NO which is converted in the atmosphere to the much more harmful NO_2 by chemical reactions involving ozone. High levels of NO_2 are a major concern in our cities because of health effects, particularly their contribution to excess deaths. These issues are addressed in Chapter 4.

NO_x and its compounds are highly reactive in the atmosphere and contribute to acid rain, eutrophication and the formation of fine particles and ground-level ozone. They are readily transported across international boundaries and so impacts are felt across entire regions. Acid rain and eutrophication lead to environmental damage in the form of loss of fish stocks, destruction of sensitive habitats and the loss of biological diversity. Fine particles and ozone cause health effects and ozone damages crops and vegetation. These issues are addressed in Chapter 5.

VOLATILE ORGANIC COMPOUNDS (VOCS)

Volatile organic compounds (VOCs) comprise the many hundreds to thousands of organic compounds that are present in the atmosphere. They range from simple hydrocarbons (such as alkanes and aromatics) to oxygenates (such as alcohols and ketones). They are pollutants in their own right because some cause cancer, such as benzene, but it is their substantial contribution to ground-level ozone formation which is of greater concern. The sources of VOCs are both man-made (vehicle exhaust, gasoline evaporative emissions and solvents) and biogenic (from trees and plants). Industry has made substantial efforts to reduce VOCs, through the development of three-way catalyst systems and evaporative canisters in gasoline vehicles and solvent reformulation including water-based paints. These issues are discussed in Chapter 5.

Individual VOCs such as benzene remain a concern because they can cause cancers. The issue of benzene as an air pollutant is now largely associated with cigarette smoking, particularly indoor air quality, as examined in Chapter 3.

METHANE (CH_4)

Methane is the largest constituent of natural gas which is a widely used fossil fuel in power stations for generating electricity, in industry for the manufacture of fertilisers and in the home for heating and cooking. Methane is emitted in large quantities by agriculture, particularly rice paddies and by animals, notably cattle. There are important natural methane sources including tropical wetlands, bogs and marshes. It is the second most important man-made greenhouse gas after carbon dioxide and is discussed in detail in Chapter 6.

OZONE (O_3)

There are no important direct emissions of ozone into the atmosphere and all the ozone present there has been formed by atmospheric chemical reactions. In the upper atmosphere, the so-called stratosphere, ozone acts as a sunscreen and filters out the harmful ultra-violet radiation from the sun. This is the stratospheric ozone layer which is essential to all life as we know it. In the lower atmosphere, particularly that portion closest to the ground, ozone is a man-made pollutant causing health effects and damage to crops and vegetation.

At ground level, ozone is a secondary pollutant formed by sunlight-driven chemical reactions involving VOCs and NO_x precursors and is transported across international boundaries by wind fields. As such, it is being addressed by all countries, which continue to reduce these two main precursor pollutants. Both chemical reactivity and meteorology are the key factors in ground-level ozone formation and mathematical models are used to predict the optimum strategy for ozone reduction in any region, as seen in Chapters 5 and 7. It is interesting to note that isoprene is one of the most reactive VOCs in ozone formation and is a major component of natural biogenic VOC emissions from trees. Strategies to control ozone formation have to take into account both man-made and biogenic VOCs in any particular region, together with NO_x emissions. It has become recognised that the formation of ozone across Europe and large parts of North America are now NO_x-limited and, consequently, further NO_x emission reductions will have the greater impact on reducing ozone levels compared with VOC reductions.

Ozone is also a man-made greenhouse gas, being the third in importance after carbon dioxide and methane. Global ozone levels are controlled by the emissions of methane, carbon monoxide, oxides of nitrogen and VOCs. Its role as a greenhouse gas is discussed in Chapter 6.

PARTICLES

The term 'particles' refers to the complex mixture of solid or liquid material that is finely dispersed in the atmosphere. It may be termed particulate matter, suspended particulate matter, aerosol or aerosol particles. Most recently, concerns have been raised about the coarse particles of size less than 10 microns in diameter (the so-called PM_{10}) and the fine particles of size less than 2.5 microns ($PM_{2.5}$). There is much current debate in the health-effects community concerning where, in terms of particle size, the toxicity resides. While this has not yet been satisfactorily resolved, it is clear that particles can be extremely damaging to human health. Smoke from wildfires, for example, can cause large increases in respiratory emergencies and hospital admissions. Open agricultural burning is an important seasonal source of particles.

$PM_{2.5}$ can be classed as primary, such as those from the tail pipe of the diesel engine, secondary inorganic aerosol (ammonium sulphates and nitrates) formed by the atmospheric chemical reactions of SO_2, NO_x and NH_3 and secondary organic aerosol formed by the atmospheric reactions of VOCs. Over the past decade, diesel engine exhaust particle emissions have been reduced by new diesel engine technology and the steady decline in emissions of sulphur dioxide and nitrogen oxides has reduced the formation of secondary inorganic particles. Road vehicles, whether petrol, diesel or electric passenger cars or whether buses and coaches, light- or heavy-goods vehicles, cause particle emissions through brake, tyre and road surface wear. Strategies for improvement of particle air quality need to consider non-exhaust vehicle emissions. The subject of secondary organic aerosols remains a key challenge. Particles and their health effects are important issues for urban air quality (Chapter 4) and regional air quality (Chapter 5).

Particles play an important role in climate processes. They scatter incoming solar radiation and control the formation of clouds and rainfall. Particles, thereby, exert a cooling influence on global climate and have, to a considerable extent, offset the global warming from the greenhouse gases. Their role in the global climate system is addressed in Chapter 6.

AMMONIA (NH_3)

Ammonia is the most important alkaline air pollutant and is widely emitted from agricultural systems which are its dominant source. Use of ammonium- or urea-based mineral fertilisers account for about one-fifth of European regional ammonia emissions. The remainder is accounted for by livestock farming and the storage and handling of animal manure.

Ammonia takes part in the atmospheric chemical reaction processes that lead to acid rain formation, the formation of fine particles and the deposition of the reduced nitrogen compounds that drive the over-supply of nutrients (eutrophication) to sensitive ecosystems and the attendant loss of biological diversity. Regional air pollution from ammonia is addressed in Chapter 5.

OTHER AIR POLLUTANTS

Toxic organic micro-pollutants (TOMPs) are produced by the incomplete combustion of fuels and classed as polycyclic aromatic hydrocarbons (PAHs), polychlorinated biphenyls (PCBs) and dioxins, all of which are emitted in very small quantities. Emissions of PAHs have been reduced substantially by three-way catalysts in gasoline exhaust systems and oxidation catalysts in diesel exhaust systems. They are still emitted from incomplete combustion in coal- and oil-fired heating systems.

Whereas the term air pollution goes way back in history, the term air quality is relatively modern. Air quality describes how polluted the air is that we breathe, and whether it is clean enough for humans and plants to remain healthy. When air quality is poor, pollutants in the air may be hazardous to people and the environment. Governments and politicians formulate plans to improve air quality.

Air quality improvement represents a huge challenge to modern society, which is the first subject of this monograph.

BIBLIOGRAPHY

Air Quality Expert Group, 2019. *Non-exhaust emissions from road traffic.* London: Air Quality Expert Group, Crown copyright.

Beddows, D.C.S., Harrison, R.M., 2021. PM10 and PM2.5 emission factors for non-exhaust particles from road vehicles: Dependence upon vehicle mass and implication for battery electric vehicles. *Atmospheric Environment* 244, 117886.

Expert Panel on Air Quality Standards, 1999. *Polycyclic aromatic hydrocarbons*. London: Department of the Environment, Transport and the Regions.

McFiggans, G., *et al.*, 2019. Secondary organic aerosol reduced by mixture of atmospheric vapours. *Nature* 565, 587–593.

Pearson, J.K., 2001. *Improving air quality: Progress and challenges for the auto industry*, Warrendale, PA: Society of Automotive Engineers, ISBN 0-7680-0236-2.

2

HISTORICAL CONTEXT

OVERVIEW

In this chapter, we trace the origins of air pollution and global warming, beginning with the last ice age, over 10,000 years ago. Indoor air quality became a concern around 2500 BC, and coal was mined by the Romans in Britain and the Rhineland to heat public baths and for the smelting of iron ores. Urban air quality soon became an issue in Greek and Roman cities. It was not until 1824 that the greenhouse effect was discovered and the role of carbon dioxide described. It was Arrhenius in 1896 who first quantified global warming from increases in carbon dioxide concentrations. In the meantime, carbon dioxide concentrations started to climb from 280 ppm (parts per million) to over 400 ppm today.

Ground-level ozone was recognised as a problem after the Second World War by Haagen-Smit in 1952 by examining the causes of Los Angeles smog. He identified how ozone was formed by the action of sunlight on nitrogen oxides and volatile organic compounds. Regional Air Quality became a serious concern and international bodies, such as the United Nations Economic Committee for Europe (UNECE), were set up to ensure co-operation between countries to tackle emissions.

It was the advent of atomic bomb testing, leading to nuclear fallout all over the world in the 1950s, which introduced scientists and governments to global air pollution. Scientific studies then led to the study of transport of carbon dioxide from industrial regions in

DOI: 10.4324/9781003293132-2

the northern hemisphere to all parts of the globe and the associated global warming. We now have the Intergovernmental Panel on Climate Change (IPCC) and the Framework Convention on Climate Change (UN FCCC) to regulate the emissions of carbon dioxide and other greenhouse gases and to monitor global warming consequences.

To set the scene for our study of the challenges brought by air pollution and global climate change, our story begins in the last ice age, which gripped Earth between 18,000 and 10,000 years ago. This ice age was marked by ice sheets covering all of Canada, the northern states of the USA, northern Britain, Germany, Poland and Russia. Glaciation in the southern hemisphere was less extensive. Because so much water was frozen in the ice sheets, sea levels were 130 metres lower than the present day, allowing *Homo sapiens* to migrate into North America, Eurasia and Australia and into Britain across the land bridge to mainland Europe. This period falls into the stone-age when the human population lived as hunter-gatherers in sparsely wooded areas, using simple wooden and stone tools. At the end of the ice age, the ice sheets started to recede and sea levels began to rise. By 10,200 years ago, the land bridge between Britain and Europe was submerged and the coast of Queensland became flooded, ultimately leading to the formation of the Great Barrier Reef. Sea level rise continued steadily, stabilising about 6000 years ago at close to pre-industrial levels. The ice sheets were replaced by steppes and prairies, boreal forests and tundra. Atmospheric concentrations of carbon dioxide began to rise as the cold oceans warmed, releasing the huge amounts they had stored during the ice age, then stabilising for a period of several thousand years.

From about 6000 years ago (4000 BC) onwards, sea levels, global surface temperatures and carbon dioxide concentrations remained constant, albeit with some minor variations. By this time, farming had already begun with the cultivation of cereal crops and the domestication of animals. Bronze was first known from around 2500 BC. Indoor air quality was evidently an issue in the poorly-ventilated and highly polluted conditions inside huts and other dwellings from cooking fires burning wood, animal dung or charcoal. Early civilisations emerged in Egypt and Mesopotamia around 3000 BC.

Cities were established to produce manufactured goods and to provide protection for their citizens. Energy was provided by wood and charcoal, although there are records of coal use dating from 3490 BC in China. The Romans mined coal in Britain and in the Rhineland to heat public baths and for the smelting of iron ores. Urban air quality in early Greek and Roman cities could sometime become poor and consequently a serious health concern. Attempts were made to keep small-scale polluting industries in the outskirts of cities. Escape to the countryside could offer some relief to asthma sufferers.

As early as 1228 AD, sea coal was being transported from the north-east of England to London for metal working and smelting and for cement manufacture. This industrial use of coal caused widespread air pollution in London and was evidently injurious to health, beginning a centuries long struggle against air pollution which is the first topic of this monograph. Coal use in the home in Britain began in the fourteenth century and expanded rapidly in the sixteenth century as supplies of wood became scarce. North American coal deposits were discovered by French explorers in New Brunswick, Canada in the 1600s and as early as 1643, coal was being shipped to the British colony in Boston.

Coal use grew rapidly with the start of the Industrial Revolution in Britain around 1750 onwards and then spread to continental Europe, the United States of America and Japan. Coal was used to power steam engines, being much cheaper and more efficient than wood, revolutionising factories, rail transport and sea transport. Later, coal was used in the iron and steel industries, for coal gas generation for home heating and for electricity generation in large coal-fired power stations. For 150 years, coal was the dominant energy supplier worldwide and even today, coal use is still significant in electricity generation despite the growth in its competitors, petroleum and natural gas.

Coal burning inevitably led to increased carbon dioxide emissions and so not long after the onset of the industrial revolution, atmospheric carbon dioxide concentrations began to rise from their pre-industrial level which had been stable for 5000 years or more. Historical carbon dioxide concentrations are available by analysing the contents of the tiny air bubbles trapped in the snow falling on the ice sheets in Antarctica. By drilling down into the ice, scientists can go back in time and follow the historical changes in the composition of the atmosphere over centuries and millennia, as shown in Figure 2.1.

First mentions of the greenhouse effect and the roles played by greenhouse gases including carbon dioxide were made by Joseph Fourier from France in 1824 and by John Tyndall from Ireland in 1861. Fourier described the natural greenhouse effect and how the atmosphere traps the heat radiating from Earth's surface, leaving Earth much warmer than it would be otherwise. Tyndall established the experimental basis for the greenhouse effect and for the roles played by water vapour and carbon dioxide in warming Earth. It was left to Svante Arrhenius from Sweden in 1896 to quantify the global temperature increase that would come from

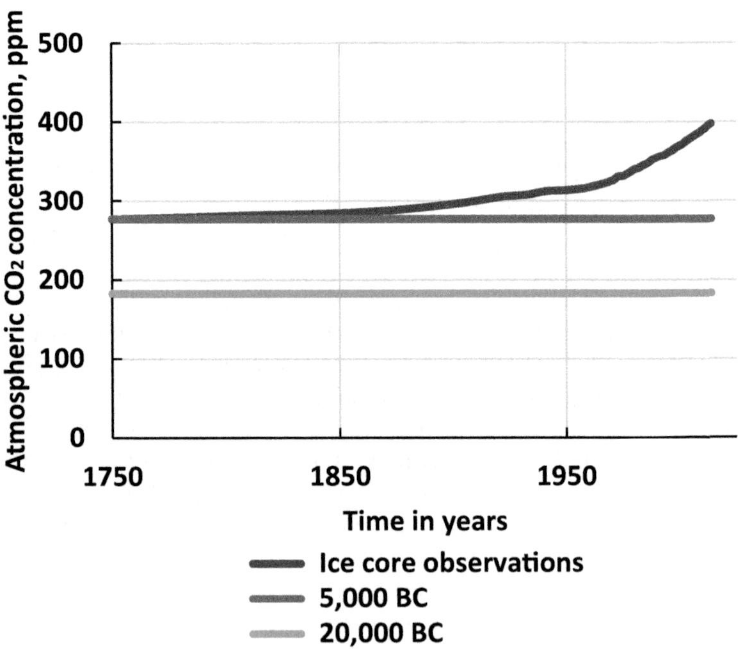

Figure 2.1 The history of the evolution of global atmospheric CO_2 levels since the Industrial Revolution as revealed from measurements in ice cores, relative to the levels found from 20,000 BC and 5000 BC onwards.

increased levels of carbon dioxide. Thus, our century long struggle began with global climate change and the role played by greenhouse gases, the second subject of our monograph.

During the nineteenth century, the levels of smoke and sulphur dioxide in our cities from coal burning began to rise rapidly. Smoke particles act as nuclei for fog droplets and so English cities became inextricably linked to fogs. It soon became evident that there were also serious health concerns associated with urban fogs in terms of illness and deaths. Occasionally, episodes of pollution would occur when stagnant weather conditions limited the natural dispersion of pollutants. During these episodes, the fogs could be quite frightening and unpleasant, causing coughing, pains in the chest and wheezing. Noticeable increases in deaths occurred during these winter episodes, notably in the Meuse Valley of Belgium during December 1930, Denora, Pennsylvania, USA in October 1948, London, UK in December 1952 and New York, USA in November 1953. In London, about 4000 excess deaths occurred during the 1952 smog episode, largely due to cardiorespiratory disease and bronchitis, with a considerable increase in deaths in younger age groups. It was important in bringing forward legislation through the Clean Air Act which paved the way for the introduction of smokeless fuels into homes and for the siting of polluting industries outside of towns and cities.

Towards the end of the Second World War, a totally new pollution phenomenon developed following the intense industrial and population expansion in the Los Angeles region of the south-west USA. It had several characteristics that made it different from the 'London-type' fogs and was termed 'Los Angeles smog' or 'photochemical smog'. It was not restricted to cold, damp winter-times and to urban centres because it enveloped both suburban and rural areas at significant downwind distances from urban centres. It was characterised by haze, eye irritation, the degradation of rubber and plastics and was driven by sunlight. It took the residents of Los Angeles by complete surprise and took many years from its initial detection to determine its causal mechanism. Haagen-Smit in 1952 described how sunlight-driven chemical reactions converted organic compounds and oxides of nitrogen, the precursors emitted largely by motor vehicles, into ozone and PAN, both unpleasant and aggressive oxidants. He was investigating the effects of photochemical smog on citrus fruit. Ozone and PAN were not emitted into the atmosphere as they are secondary pollutants but were formed there by chemical reactions involving the emitted precursors. Photochemical smog was therefore inextricably linked to the rise of the motor vehicle and a shift in concern from the use of coal to the use of oil, thereby bringing into focus a new form of pollution: regional air pollution. Subsequently, photochemical smog has been detected throughout North America, Europe and Asia.

Regional air pollution problems extend over considerable distances, many tens to hundreds and up to thousands of kilometres downwind of sources, because the atmospheric chemical reactions that cause them take a considerable amount of time to occur. These chemical reactions generate aggressive and acidic secondary pollutants which can cause damage to human health, crops and vegetation, to materials and to sensitive ecosystems. Examples of regional air pollution include Los Angeles smog, acid rain damage to forests, acidification of soils and surface waters with the loss of fish stocks, corrosion of materials and buildings with the loss of cultural heritage, and the eutrophication or over-fertilisation of sensitive ecosystems with the loss of biological species diversity. In all these cases, the impacts of regional air pollution were not anticipated in advance and took communities by surprise. They all took intensive scientific research to come up with the basic scientific understanding to develop control strategies and policies. Because regional air pollution problems are often found at considerable distances downwind of their precursor emission sources, problems can be transported over state or national boundaries. There are therefore often trans-boundary issues associated with regional air pollution problems. Such issues have been particularly important in Europe and have led to international initiatives, treaties and conventions to pool scientific research, to harmonise emission inventories, to coordinate emission reductions and so to improve regional air quality on a cooperative and coordinated basis. An important example of international cooperative action has been UNECE and its Convention on Long-Range Transboundary Air Pollution (CLRTAP).

At the height of the Cold War between 1952 and 1958, the USA, UK and the USSR began testing atomic weapons in the atmosphere at the Marshall Islands and Christmas Island in the remote regions of the Pacific Ocean and at Novaya Zemlya in the Siberian Arctic. Early studies of fission products showed that the detonations produced airborne radioactive material in the form of particles. The particles injected into the lower atmosphere were deposited quickly, within 20–30 days. However, some fission products were directly injected into the stratosphere which acted as a global reservoir which spread out over both hemispheres. Every spring-time, air motions would bring stratospheric air laden with fission products into the lower atmosphere, transporting nuclear fall-out to all corners of the globe. This was our first glimpse of a new form of air pollution: global air pollution. In the face of mounting international condemnation, atmospheric testing of nuclear weapons finally ceased under the terms of the United Nations (UN) Partial Test Ban Treaty of 1963.

During the International Geophysical Year of 1957, Charles Keeling from the USA began measuring the levels of carbon dioxide in the atmosphere at the top of Mauna Loa, Hawaii and at the South Pole. It soon became apparent that the carbon dioxide observations at the South Pole were increasing in step with those being made at Mauna Loa in the northern hemisphere. There must be atmospheric motions able to transport the long-lived carbon dioxide from the industrial sources in the northern hemisphere to the remote South Pole. Carbon dioxide and the greenhouse effect it drives, are evidently another form of global air pollution. It took many years for the real significance of Charles Keeling's South Pole observations and the early discoveries of Fourier, Tyndall and Arrhenius to be recognised. Ultimately, they led to the setting up of the Intergovernmental Panel on Climate Change (IPCC) in 1988, the United Nations Framework Convention on Climate Change (UNFCCC) in 1994 and its Kyoto Protocol in 1997.

A third example of global air pollution involves the destruction of the stratospheric ozone layer which protects Earth's surface from the damaging ultra-violet radiation from the sun. Concerns were first voiced in the early 1970s in the USA about the exhaust emissions from high-flying supersonic Anglo-French Concorde aircraft destroying the ozone layer. However, because of the limited number of commercial supersonic aircraft planned, the threat was finally considered to be small. However, at this time research indicated that man-made chlorofluorocarbons (CFCs) widely used as aerosol propellants and refrigerants could potentially catalyse the destruction of ozone in the upper stratosphere. The proposition was that the CFCs had no known removal processes in the lower atmosphere and so would be transported into the upper stratosphere where they could be broken down by the ultra-violet radiation from the sun. This would liberate highly reactive chlorine atoms which would catalyse the destruction of ozone, exposing Earth's surface to damaging ultra-violet radiation, leading to the increased incidence of skin cancer. These concerns were sufficiently strong for the nations to come together and agree the United Nations Vienna Convention for the Protection of the Ozone Layer in 1985 and to draw up the Montreal Protocol to that Convention in 1987.

In 1985, Joe Farman of the British Antarctic Survey published evidence of the appearance of a 'hole' in the stratospheric ozone layer over Antarctica during each spring since his measurements began during the 1970s. A combination of aircraft flights over Antarctica and laboratory investigations subsequently revealed the causal mechanism behind the formation of the Antarctic ozone 'hole' to involve the exceedingly low temperatures found in the stratosphere over the Antarctic, polar stratospheric clouds and the chlorine liberated from the destruction of man-made CFCs in the upper stratosphere. Under the Montreal Protocol, CFCs were phased out and were replaced as refrigerants and foam blowing agents by hydrochlorofluorocarbons (HCFCs). But as the Antarctic ozone 'hole' grew larger in size each spring and 'holes' began to appear each spring in the Arctic ozone layer, these also had to be phased out. Hydrofluorocarbons (HFCs) were selected as replacement refrigerants and air-conditioning fluids, but they are potent greenhouse gases and are thus adding to global climate change.

Looking back over Man's history up until the end of the twentieth century, four types of pollution have been identified: indoor, urban, regional and global and these form the subjects of the next chapters of this monograph. We discuss how these four types of pollution problems are manifested in the contemporary world and the scientific basis of our understanding of each. Armed with this, we can begin to address the challenges faced by improving indoor, urban, regional and global air quality both singly and together in the years ahead.

BIBLIOGRAPHY

Brimblecombe, P., 1999. Air pollution and health history. In S.T. Holgate, J.M. Samet, H.S. Koren, R.L. Maynard (eds), *Air pollution and health*, pp. 5–18. London. Academic Press.

Clapp, B.W., 1994. *An environmental history of Britain since the Industrial Revolution*. Harlow: Longman.

Fowler, D., *et al.*, 2020. A chronology of global air quality. *Philosophical Transactions of the Royal Society* A378, 20190314.

Fuller, G., 2019. *The invisible killer: The rising global threat of air pollution – and how we can fight back*. London: Melville House.

Haagen-Smit, A.J., Bradley, C.E., Fox, M.M., 1953. Ozone formation and the photochemical oxidation of organic substances. *Industrial and Engineering Chemistry* 45, 2086.

Junge, C.E., 1963. *Air chemistry and radioactivity*. New York: Academic Press.

Leighton, P.A., 1961. *Photochemistry of air pollution*. New York: Academic Press.

3

INDOOR AIR POLLUTION

OVERVIEW

Given that humans spend 80% of their lives indoors, good indoor air quality is of paramount importance. In this chapter, we examine the indoor air pollutants of most concern and the need for indoor air quality standards. $PM_{2.5}$ particles are of greatest concern, followed by VOCs, especially those which are carcinogenic. We suggest that indoor air quality standards for $PM_{2.5}$ and toxic VOCs should mirror those for outdoor air quality. We rely on the chemical and solvent businesses to define a labelling system for other VOCs rather than have unnecessary legislation imposed.

Indoor air pollution consists of gases or particulates which cause acute or long-term health effects when we breathe them in. The important fact is that we spend 80% or more of our time indoors which makes good indoor air quality a prime consideration in our daily lives. It is an important source of exposure to air pollution because we spend so much of our time indoors, at home, or at work, or at school, or when we go out shopping or to the cinema, for example. Indoor air quality consists of two main components, the air quality from outside which enters the building and the air quality generated within the building. To understand indoor air pollution, we must deal with the accumulation of pollutants within the home or inside buildings, caused by poor ventilation and inadequate air exchange.

DOI: 10.4324/9781003293132-3

Indoor air pollution is one of the earliest forms of air pollution as we have explained in Chapter 2. It is historically associated with cooking and this association is still important today. The majority of households in Africa, particularly sub-Saharan Africa, Latin America and south Asia, use solid fuels or simple stoves which release most of the smoke into the home. The resulting indoor pollution is a major threat to health particularly for women and young children who may spend hours close to the fire. Biomass fires, particularly in poor households, can lead to chronic obstructive pulmonary disease (COPD), respiratory infections, pneumonia, stroke, heart disease and lung cancer which can result in premature death. It is an important risk factor for premature death in the global burden of disease, with indoor air pollution lying fourth in importance below high blood pressure, smoking and high blood sugar. It is estimated that indoor air pollution causes between 1.6 and 4.3 million deaths each year across the globe.

Death rates from indoor air pollution are highest in low-income countries and have been declining in almost every country in the world since 1990. This decline is driven by increasing access to clean fuels. As household incomes rise, traditional solid fuels such as crop waste, dung and firewood, give way to charcoal and coal. With higher household incomes, solid fuels give way to clean fuels such as ethanol, LPG, natural gas and electricity. Solar cooking stoves can be useful where conditions allow. The world is making progress in shifting to more modern energy sources for cooking and this has been particularly noticeable in south Asia.

Indoor air pollution from burning fuels for cooking is generally less of a concern in modern homes where heating systems, damp, household cleaning and some building materials are much more of a concern. Sometimes, better ventilation can improve indoor air quality but at the expense of deteriorating energy efficiency. Indoor air quality can also be affected adversely by outdoor air quality, particularly during air pollution episodes. Office equipment can be important sources of air pollution at work. Workplaces have their own particular issues, depending on how polluting their operations are within the indoor environment.

Cooking and heating may give rise to particles and gases such as carbon monoxide, oxides of nitrogen and sulphur dioxide. Smoke and particles from cooking have been an issue for indoor air quality since pre-history. Indoor exposure to carbon monoxide is a major cause of death, either intentionally or unintentionally. Suicides by carbon monoxide poisoning have declined as town gas, which is largely carbon monoxide, has been replaced by natural gas, which contains virtually none. Carbon monoxide is emitted by incomplete combustion of charcoal, coal and coke, and from poorly serviced cookers, heaters, boilers and other appliances burning natural gas or paraffin. This is a major contributor to accidental exposure which may lead to unintentional death. All buildings should be fitted with CO sensors as a matter of necessity.

Burning coal or wood in an open grate or stove gives rise to smoke emissions not only indoors but also outdoors in the local neighbourhood. Wood-burning stoves are becoming a growing concern in towns and cities and smoke control areas are being established to control the type of fuels being burnt to only dry, unpainted or untreated wood, restricting the burning of household rubbish and packaging and regulating the types of stoves that can be used to the more modern examples.

Cooking with natural gas produces carbon monoxide, water vapour and oxides of nitrogen and has been associated with respiratory illnesses in children. However, there are some key uncertainties in our understanding of the effects of cooking with natural gas on respiratory illness. Whereas some studies of children have shown an association between gas cooking and increased respiratory illness, others have demonstrated no association between gas cooking or measured NO_2 concentrations and illness. The inconsistencies in these studies can be explained by the lack of significant statistics because of the small sample sizes, the effects of other combustion products, confounding by outdoor air pollution from nearby traffic and inadequate or insensitive measures of health effects. Gas cooking does not just produce elevated NO_2 levels alone but leads to increased emissions of VOCs, CO, SO_2, water vapour, particles, CO_2 and other oxides of nitrogen. These issues lead to real uncertainties in studies of the health effects from cooking with natural gas – nevertheless a small effect cannot be ruled out. Ventilating the kitchen or using a cooker hood when operating gas stoves may reduce the risk of respiratory symptoms.

A wide range of household chemicals are used to clean and decorate our homes and some are associated with breathing difficulties and setting off allergies. Some of these chemicals may contain volatile organic compounds (VOCs) which evaporate when stored or used in the home. In some cleaning products, the VOCs are included as fragrances such as lemon or pine, merely to improve their attractiveness to customers. In other products such as paints, glues, varnishes and spray polishes, the VOC is essential to the correct effectiveness of the product. It is important to ventilate your home when cleaning or decorating using furniture polishes, carpet cleaners, air fresheners, varnishes and glues.

As we have described, the main sources of pollution generated indoors are VOCs associated with smoking, cooking and heating, and solvents used in cleaning, cosmetics and painting. Benzene from smoking has been substantially reduced in workplaces, but less so in private homes. Polycyclic aromatic hydrocarbons (PAHs) are formed by thermal degradation processes, usually incomplete combustion from coal and oil-fired heating. There is legislation on PAHs in some European countries mainly because of their link to cancer formation.

The field of indoor air quality is gathering momentum, with compounds associated with humans accounting for over 40% of indoor VOC concentrations. Microbiological activity (e.g. mould in buildings) produce VOCs. Solvent industries have reduced the VOC content in their products, for example through the extensive use of water-based paints, yet some indoor VOCs remain a concern. VOC emission rates from non-aerosol personal care products (PCPs) including limonene used in cleaning fluids are a concern in indoor atmospheres. The use of ethanol in cosmetic products and disinfectants including hand gel is substantial. However, ethanol is classed as a skin irritant and not a pollutant. The European Solvents Industry Group is liaising with academia and research institutes with a view to improving indoor air quality linked to solvent usage, in particular semi-volatile organic compounds (SVOCs). It is most likely that this work will lead ultimately to labelling schemes for VOCs and SVOCs.

Formaldehyde and its compounds, which cause irritation of the eyes and upper airways, can be found in building materials, furniture, carpets and upholstery. Chipboard (particle board), for example, is a common building construction material made of wood shavings held together with urea-formaldehyde glue. There is a continuous release into the home of formaldehyde from urea-formaldehyde building materials.

Fragrances, such as pine and citrus, contain reactive alkenes and monoterpenes which can react in the indoor environment with ozone from the outdoor environment, sometimes forming new species such as secondary organic aerosol, organic nitrates and carbonyls, some of which are known to be irritants and some are carcinogens.

Allergic reactions are sometimes triggered by indoor air pollution. Allergies to dust mites, mould spores and pets are common with asthma sufferers and may be exacerbated indoors. Pets shed dander and their saliva or urine can be allergenic. In cold climates, damp in homes is a problem because it leads to condensation which encourages moulds and fungi to grow, giving off microbial organic compounds.

Outdoor air enters homes through infiltration through windows, doors and other openings in the building and depends on the type of windows and the frequency of their opening. Ventilation systems also bring outdoor air into the home. Infiltration and ventilation may bring outdoor air pollutants indoors and this is particularly important for homes in towns and cities, particularly those adjacent to busy roads. Outdoor sources of nitrogen dioxide, particles and ozone are of particular importance in any consideration of indoor air quality. Ozone in indoor air quality is of less concern because it degenerates on solid surfaces which is why asthmatic persons are advised to stay indoors when levels are high outdoors. The main pollutant from outdoors of concern indoors, is $PM_{2.5}$, fine particles of size less than 2.5 microns in diameter which can penetrate the lung. Generally speaking, indoor levels may be up to one half of outdoor levels depending on the pollutant and the infiltration and ventilation rates.

INDOOR AIR QUALITY MODELS

In Appendix 1, we detail indoor air quality models and use them to examine three specific situations. We summarise these three applications of our indoor air quality model below.

DECORATING AN INDOOR SPACE

(APPENDIX 1, EXERCISE A1.1)

Here we envisage decorating an indoor space of 24 m^3 by applying decorative solvent-based paint at a rate of 1 litre per hour. We show quantitatively the influence of air ventilation rates on indoor paint concentrations and how long these concentrations remain after painting has ceased.

A POLLUTANT WITH BOTH INDOOR AND OUTDOOR SOURCES – ALBERT HALL

(APPENDIX 1, EXERCISE A1.2)

In this illustrative example, we answer the question about ventilation requirements for a concert venue such as the Albert Hall in London. Assuming that the Hall seats 5000 people and has a volume of 100,000 m^3, we can estimate the ventilation rate requirements to maintain the carbon dioxide concentrations below 1000 ppm, the level at which the audience will start to feel uncomfortable. The model shows that a complete air change of the internal volume is required every 40 minutes.

INDOOR PM LEVELS CLOSE TO A BUSY ROAD

(APPENDIX 1, EXERCISE A1.3)

In this example, we use the time-dependent indoor air quality model to describe indoor PM levels close to a busy road. In Figure 3.1, we illustrate the numerical solution of the time-dependent indoor air quality model for the PM concentrations inside a home.

The thick line shows an illustrative PM air pollution episode in an urban area. The peak in outdoor PM in this diagram shown as a thick solid line may, for example, represent the morning rush-hour and extends over a four-hour period. The thin solid line represents the simple indoor air pollution model indoor PM concentrations that result from modest infiltration and ventilation rates. The magnitude of the indoor PM peak is less than half the outdoor PM peak, 42 $\mu g\ m^{-3}$ instead of 100 $\mu g\ m^{-3}$, and is delayed by nearly an hour. If the ventilation rate is increased (dotted line), the magnitude of the PM peak increases by about 16 $\mu g\ m^{-3}$ and it occurs about twenty minutes sooner. However, the pollution is dissipated more quickly as the home is better

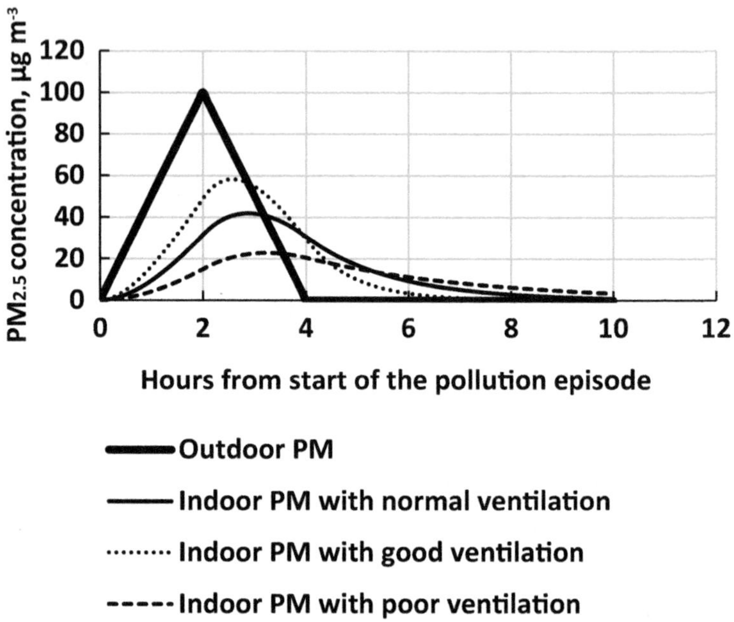

Figure 3.1 Diagram illustrating the relationship between indoor and outdoor air quality for PM using a simple indoor air quality model.

ventilated. In contrast, decreasing the ventilation rate (dashed line) decreases the magnitude of the PM peak by 20 $\mu g\ m^{-3}$ at the expense of delaying the timing of the maximum and extending time period of the exposure to the PM. Indoor air quality is thus strongly influenced by outdoor air quality, but the relationship is not straightforward, being dependent on the specific pollutant and infiltration and ventilation rates. This is why we need indoor air quality models to understand the relationships between outdoor and indoor air quality.

$PM_{2.5}$ has a high penetration factor in indoor buildings, and given this parameter includes secondary particles such as sulphates and nitrates as well as elemental carbon and organic carbon, we can identify $PM_{2.5}$ as one of the most important indoor air pollutants. Given that little $PM_{2.5}$ is generated within buildings in modern times, except where building improvements are taking place, the control of outdoor $PM_{2.5}$ is even more essential.

LEGISLATION IS A NECESSITY – BUT WHICH POLLUTANTS?

It is recognised that we spend most of our time indoors, and yet, there is little legislation on indoor air quality. It is clear that $PM_{2.5}$ remains the greatest challenge in indoor air quality, and a clear standard is needed for $PM_{2.5}$ indoors with regular monitoring. We suggest the standard for $PM_{2.5}$ should be the 2021 WHO Guideline for outdoor $PM_{2.5}$ of 5 $\mu g\ m^{-3}$ annual mean and 15 $\mu g\ m^{-3}$ 24-hour mean. With respect to indoor VOCs, those which are carcinogenic, such as benzene and PAHs, require legislation and monitoring, and these limits should be set at the same levels as outdoor air quality standards, initially. Other indoor VOCs and SVOCs are being studied by both industry and institutions and a good first step would be to publish labelling schemes for each chemical, rather than have legislators imposing limits for individual chemicals. Let us allow the labelling schemes to lead us to a gradual reduction of the more sensitive VOC products for indoor air quality.

BIBLIOGRAPHY

Beko, G., Carslaw, N., Fauser, P., Kauneliene, V., Nehr, S., Philips, G., Saraga, D., Schoemaecker, C., Wierzbacka, A., Querol, X., 2020. The past, present and future of indoor air chemistry. *Indoor Air* 30, 373–376.

European Solvents Interest Group and European Solvents VOC Co-ordination Group, 2017. *VOCs and indoor air quality*. Technical paper. Brussels, Belgium: European Solvents Interest Group and European Solvents VOC Co-ordination Group. www.esig.org/wp-content/uploads/2018/03/201709_Technical_Guidance_Paper-ESVOC-Indoor_Air_Quality_Final-1.pdf.

World Health Organization, 2021. *Air quality guidelines: Global update 2021*. Copenhagen, Denmark: Regional Office for Europe of the World Health Organization.

Yeoman, A.M., Shaw, M., Carslaw, N., Murrells, T., Passant, N., Lewis, A.C., 2020. Simplified speciation and atmospheric volatile organic compound emission rates from non-aerosol personal care products. *Indoor Air* 30, 459–472.

4

URBAN AIR QUALITY

OVERVIEW

In this chapter we examine the air pollutant emissions and atmospheric dispersion processes which drive urban air quality. Road transport is a major source of urban air pollution but there are many other stationary sources ranging from industrial plants and domestic machinery. The three main urban air pollutants are nitrogen oxides (NO_x), ozone (O_3) and particulate matter (PM), the latter two being partly transported from other regions and countries. Air quality models are used, not only to predict air pollutant concentrations but also to assess the most cost- effective solutions to improve urban air quality.

We concentrate on how pollution from road transport can be reduced over the next two decades in Europe and Asia. A new win–win fuel for heavy duty vehicles will be hydrogen which can be readily manufactured from renewable energy sources. Urban air pollution will become a problem of the past by 2050 with zero emissions road transport and greater use of rail transport.

Urban air quality is the result of the effect of two opposing forces. Air pollutant emissions raise pollutant concentrations and atmospheric dispersion processes act to reduce them. Episodes of high pollution can thus be caused by emission peaks perhaps due to traffic jams, morning and evening rush hours and traffic accidents under normal dispersion conditions, or normal emissions during weather conditions

DOI: 10.4324/9781003293132-4

with poor dispersion. Meteorology controls the extent to which pollutants are mixed vertically upwards from their ground-level sources and the extent to which they are transported out of the urban area by the wind. During winter, stagnant conditions can sometimes occur and last for several days. During these calm conditions, wind speeds fall and vertical mixing is severely limited by low level inversions which act as a lid on the urban atmosphere. Under these conditions, pollutant concentrations may build up and air quality rapidly deteriorates, threatening widespread health effects. Some of these episodes have been reviewed in Chapter 2.

Following the shift in urban areas away from coal burning in homes to natural gas during the 1960s and 1970s, urban air quality ceased to be of public concern. However, as time has passed, traffic has taken over as the main source of urban pollution in our towns and cities. When the same calm, poor dispersion conditions, for example, returned to London in December 1991 as occurred during the London smog of 1952, elevated levels of traffic-generated pollutants such as nitrogen dioxide (NO_2), PM, carbon monoxide and carcinogenic hydrocarbons were reported, while smoke and SO_2 levels were relatively low. There was an increase in mortality and hospital admissions associated with this episode. This story has been repeated in all the major cities of western Europe, North America and Asia.

Traffic is a huge source of air pollutants in urban areas because traffic pollutants are emitted close to the ground and do not benefit from good dispersion conditions to the same extent as industrial pollutants, emitted from chimney stacks. However, this does not mean that traffic is the sole source of urban pollution. There are other urban sources such as railways, airports, oil and natural gas combustion in commercial and public buildings, district heating systems and small-scale combined heat and power systems. Also, pollutants such as ozone and particles are transported into the urban environment from the surrounding regions and sometimes from neighbouring countries.

Emissions from road traffic heavily dominate over other sources in urban areas. The major categories of motor vehicles include passenger cars, taxis, public service vehicles, buses and coaches, vans, trucks, heavy goods vehicles, motorcycles, mopeds and scooters. Passenger cars can be powered by gasoline or diesel engines, or, in some countries (such as France, Netherlands and Italy), by liquefied petroleum gases (LPG). The share of the passenger car market being taken up by diesel-powered vehicles has been increasing and in many European countries such as France, this share has increased to over one half. In North America and Asia, the vast majority of passenger cars have gasoline engines. Taxis in Europe are generally diesel-powered, as are public service vehicles, buses and coaches, vans, lorries and heavy goods vehicles. Two-wheeled motorcycles, mopeds and scooters are predominately gasoline-powered. Whether urban vehicles employ gasoline or diesel engines potentially makes a large difference to their emissions of nitrogen oxides, hydrocarbons, carbon monoxide and particles, but little difference to carbon dioxide emissions. It also matters how old or new each vehicle is and whether it has emission controls fitted and working correctly. Latest technology (EURO-4) gasoline and diesel (EURO-6) passenger cars have similarly good emission performance and EURO-7 norms being introduced in 2021 will further improve new vehicle emissions.

Air pollution from urban traffic typically shows a distinct diurnal cycle with morning and evening rush hour peaks caused by high volumes of traffic and high levels of congestion. Pollutant concentrations are generally lowest during the early morning when traffic volumes are at a minimum. Off-peak levels during the late morning and early afternoon are somewhat lower than the morning and evening peaks.

Concentrations of traffic pollutants decline with distance away from busy roads. Where the road is running through suburbs or parkland, concentrations fall by one half roughly 30 metres from the road edge, by a further one quarter at 120 metres from the road edge and become undetectable at greater distances. This is illustrated in Figure 4.1, where PM concentrations downwind of a busy road are calculated using a simple urban air pollution model, see Appendix 2. However, where a busy road is in a street canyon, dispersion is strongly inhibited by the presence of buildings on either side of the road. Traffic pollutant concentrations can rise dramatically and show markedly different spatial distributions to those from an unobstructed highway. Field studies and wind tunnel studies have been used to understand the processes that occur in canyon streets. The most characteristic feature of street canyons is where the direction of the

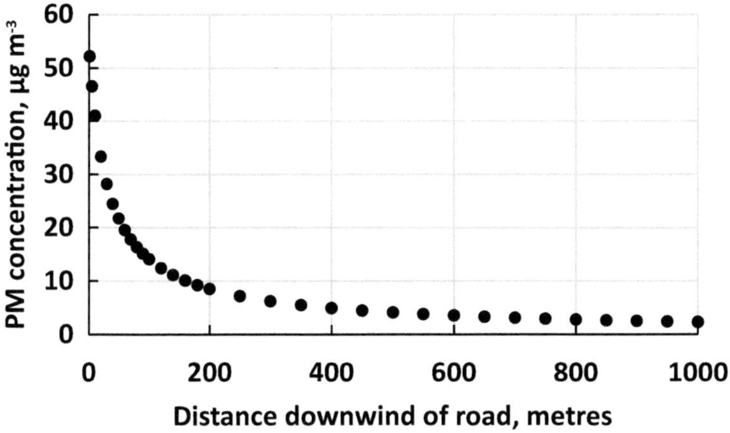

Figure 4.1 PM concentrations downwind from a busy road showing how concentrations fall off rapidly with distance.

wind at street level is opposite to the wind above the roof-tops. Understanding these microscale processes is essential to the assessment of the exposure of the urban population living in the vicinity of busy roads.

Most large cities have well established air pollution monitoring networks to provide continuous observations of urban air quality which can be directly compared against internationally accepted air quality criteria. In Europe, North America and Asia there is much public concern about levels of NO_2, PM and ozone (O_3). In Europe, for example, there are many urban areas where compliance with the EU Air Quality Directives is proving difficult for both NO_2 and PM. These urban areas include Athens, Krakow, Lisbon, London, Milan, Paris, Sofia and Stuttgart. Monitoring network data when coupled with health effects monitoring can provide robust and valuable epidemiological data for both episode conditions and long-term studies.

Air pollutant monitoring data have been used extensively to learn about the behaviour and origins of pollutants in the urban environment. As an example, we present an analysis in Figure 4.2 of the monitoring data for NO_x and O_3 at a monitoring station sited at the kerbside of a busy road in the centre of London during

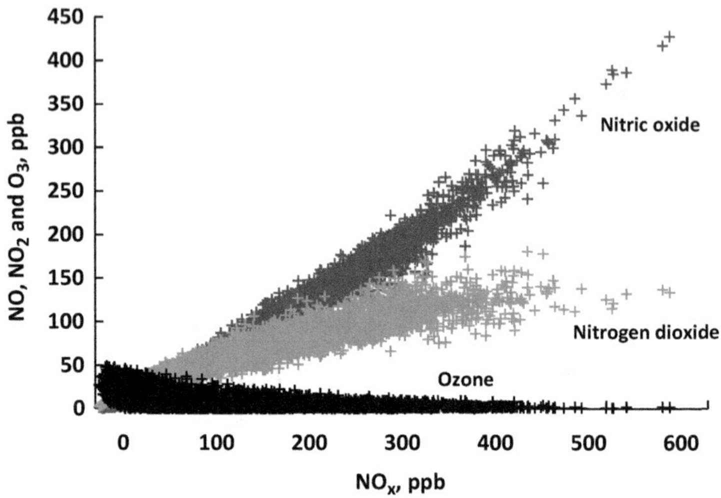

Figure 4.2 Hourly air quality monitoring data for the Marylebone Road, London kerbside monitoring station for 2010 for NO, NO_2, O_3 plotted out against NO_x.

2010. Much of the NO_x emitted by road traffic in London is emitted as NO. Some of this NO is oxidised to NO_2 by the O_3 that is transported into the centre of London by the wind. This chemical reaction depletes the O_3 and produces NO_2 which is much more harmful to human health than the emitted NO. In Figure 4.2, we plot the simultaneous hourly values of NO, NO_2 and O_3 from the monitoring network in a scatter plot against the values of $NO_x = NO + NO_2$. As NO_x levels increase from left to right, we move from the relatively unpolluted early morning conditions through to the grossly polluted conditions associated with gridlock. O_3 levels (shown as black crosses) decline as we move from unpolluted to polluted conditions due to depletion by reaction with NO. Eventually all the O_3 imported into London is removed and at the right-hand side of the figure we see that all the ozone levels are zero. NO_2 levels (shown as grey crosses) are close to zero in the unpolluted conditions at the left-hand side of the diagram and increase steadily as pollution levels rise, eventually levelling out at high NO_x levels. The rise in NO_2 initially follows the decline in O_3 due to chemical formation but increases above the 50 parts per billion (ppb) level from the imported O_3 (where 1 ppb represents a mixing ratio of 10^{-9} mole per mole air). This is because the traffic itself is a source of NO_2. Under the most polluted conditions at the right-hand side of the diagram, NO_2 levels are about one quarter of the NO_x levels so that the traffic must be emitting about one fifth of the NO_x as NO_2 directly, with the remainder as NO, making allowance for the NO_2 formed from O_3. Diesel buses, taxis and light goods vehicles are responsible for this direct emission of NO_2. The upshot of this is that long-term mean NO_2 levels are controlled by the imported O_3 and so are beyond the immediate control of the urban area, whereas the peak short-term levels are controlled by the direct emission of NO_2 and so respond to traffic management and vehicle emission reductions.

Air quality monitoring data are reported to the European Environment Agency (EEA) from more than 4000 monitoring stations across Europe. Over the years, these data have shown a slow improvement in urban air quality but concentrations are still above health effects thresholds. Levels of NO_2 and PM are of particular concern for causing heart disease and respiratory problems, leading to premature death. Across Europe, 412,000 premature deaths in 2016 were attributed to $PM_{2.5}$ exposure and 71,000 to NO_2 exposure. The largest number of premature deaths due to $PM_{2.5}$ was found in Germany, Italy and Poland but on a per capita basis, in Kosovo, Serbia, Albania and North Macedonia. For NO_2, the countries with the largest numbers of deaths were Italy, Germany and the UK, but on a per capita basis they were Greece, Italy, Serbia, Cyprus and the UK.

While monitoring networks can answer many of the policy-relevant questions concerning urban air quality, there are some questions that require the application of modelling tools to bring forward answers. These questions may address the origins of any air quality exceedances in terms of which are the most important emission sources that contribute to the exceedance of air quality criteria. It may be important to know whether sources outside of the urban area also contribute to exceedances and this is most important for PM. Then there are questions concerning future air quality exceedances and whether exceedances will cease naturally because of actions and control measures that have already been planned. It may be that additional measures beyond business-as-usual will be required if air quality targets are to be met. Equally well, it may be interesting to know what could be achieved if all feasible emission reductions and control measures were implemented. All these strategies require analysis and assessment to identify the costs involved and whether the benefits of any air quality improvement outweigh the costs. These questions can be answered with urban air quality models.

A range of urban air quality modelling tools is available to policy-makers to analyse and assess control strategies designed to improve air quality. Examples of urban air quality models are described in Appendix 2. Urban air quality models share a number of common features. They must include comprehensive emission inventories for the major traffic and stationary sources and for the major urban air pollutants: NO_x and PM. Meteorological data are required to describe how pollutants are dispersed away from their sources, how the pollutants from the individual sources merge into an urban plume and how the urban plume is then transported out of the city centre and over the suburbs. Meteorological data describe the import of O_3 from the background atmosphere and how the urban area is ventilated and pollutants are transported into downwind regions. Air quality models should include a chemistry module to describe how emitted NO is oxidised by imported O_3 to form NO_2 and the transformations that make secondary PM components from primary PM precursors. Policy-makers can reasonably expect that urban air quality models should be able to reproduce real world behaviour as revealed by urban air quality monitoring, as exemplified in the above diagram showing NO–NO_2–O_3–NO_x relationships, for example. Hence, a meaningful comparison between model predictions and observations is an essential prerequisite for building confidence in air quality models and for supporting their role in designing strategies for urban air quality improvement.

Despite five decades of policy action to control motor vehicle emissions, urban air quality in Europe is still a major issue. Levels of NO_2 and PM in our cities regularly breach EU Air Quality Directives leading to thousands of premature deaths across Europe. With the recent decisions to phase out internal combustion engines in EU passenger cars by 2030 in favour of electric vehicles, the prospects of achieving acceptable air quality have improved significantly. This still leaves diesel vans, light goods vehicles, buses, coaches and heavy goods vehicles as major sources of urban PM and NO_x which require further action.

THE WAY AHEAD

We now examine urban transport in detail and, in particular, how this will change in the next two decades. Urban air pollution fell by 30% in the first three months of the lockdown in 2020 due to the Covid pandemic. The challenge is to return to those low levels of air pollution by modernisation of urban transport.

The transformation of passenger car road transport from gasoline and diesel fuel to hybrid and electric engines has already begun, and requires a concerted effort especially with respect to the replacement of old vehicles in Europe.

EUROPEAN PASSENGER CARS

European passenger cars contribute nearly 20% of all CO_2, and diesel cars and heavy-duty vehicles contribute to 50% of all NO_x generated in Europe. The breakdown in car numbers in Europe in 2020 was 52% gasoline, 2% electric with the remainder almost all diesel. In Eastern Europe, there are a high percentage of passenger cars over 20 years old, with over 36% of cars in Poland over 20 years. This is the real opportunity for governments and the EC to provide generous scrappage allowances to support people who can exchange their cars for new modern technology versions, particularly electric vehicles.

Total emissions from each type of vehicle comprises emissions generated in use, plus production plus disposal, which only gives electric vehicles a real advantage if electrical power in production, use and disposal is generated by renewable energy sources.

Motor manufacturers have responded to the challenge of constructing new electric and hybrid vehicles with all production of diesel and gasoline cars being phased out by 2030 at the latest, and one manufacturer is stopping the production of gasoline cars by 2022. Indeed, NO_x reduction will be dramatic as diesel passenger car manufacture is phased out.

Average passenger car CO_2 emissions, currently at 118.5 gm km^{-1} is being reduced to less than 95gm km^{-1} in 2021 with further reductions planned in 2025 and 2030.

The scrappage of older vehicles is essential, and governments should identify cars with high emissions levels on a regular basis and encourage motor manufacturers to make generous contributions towards the purchase of new electric vehicles.

European statistics reveal that in 2018 there were 47 million passenger cars in Germany, 39 million in Italy, 32 million in both France and the UK and 23 million in Poland. Are all these cars necessary? A detailed inventory of all passenger cars including their age and emissions category is available for each country and the emissions reductions can be assessed for any government strategy. Older higher emissions cars can be phased out. It is interesting to note that during the Covid pandemic, many families had two or more cars standing in their drives. Are all these cars essential in normal times? Can we make better use of public transport? Reducing vehicle emissions involves changed and improved lifestyles. By 2030, Europe can reduce both the number of cars and total emissions substantially, and by 2040 at the latest we will see real blue skies in our cities and urban areas as can be seen in the mountains of Switzerland.

OTHER EUROPEAN TRANSPORT

European transport currently accounts for 30% of EU CO_2 emissions. The EU objective is to reduce CO_2 emissions by 30% in 2030 relative to 2019 levels. Whereas passenger cars now account for 60% of all transport CO_2 emissions, heavy duty trucks account for 26%, light duty vehicles 12% and railways for 0.5%. While new technology for truck motive power is being developed, there is a window of opportunity to build up goods traffic on European railways and so reduce road transport emissions rapidly. Yet, a breakthrough in replacing the diesel engine in heavy duty vehicles and buses by clean automotive power, using for example hydrogen, is already appearing.

ROAD TRANSPORT IN ASIA

China had a total of 258 million passenger cars in 2019 of which 13% were hybrid petrol cars and the remainder gasoline. Given the fact that most passenger vehicle journeys are start-stop in congested cities, the challenge is to replace petrol cars by both hybrid and electric vehicles now! China has the will to respond to rapid change and with a rapid expanding economy, gasoline passenger cars will be a distant memory within a decade.

Asia is following the lead of Europe in replacing their automotive fleets with electric cars, manufactured in their own countries. It could even be achieved more quickly than Europe.

THE CHALLENGE OF HYDROGEN IN FUTURE TRANSPORT

Hydrogen (H_2) is a clean fuel releasing energy and water on combustion, with no pollutant emissions. It is already being used in newly manufactured lorries and could well be used in ships and central heating as a replacement for natural gas. Just as oil was once termed black gold, H_2 is now the fuel of the future.

Manufacture of H_2 is by energy intensive processes at present. 'Grey' hydrogen is made by steam methane reformation, generating much CO_2. 'Blue' H_2 is made by the same process but the carbon is captured and buried underground. 'Green' H_2 uses renewable energy sources to generate H_2 by electrolysis of water. This last process is expensive at present, but this is reducing in cost as wind and solar power costs fall rapidly. Energy companies are proactive on H_2 manufacture and storage, trucks are being manufactured using hydrogen fuel cells. Government encouragement and investment is already showing here.

Heavy duty vehicles will have a long range between refuelling and time to refuel will be short. Diesel engines in trucks and buses will be phased out in the next two decades. NO_x and CO_2 emissions from road transport are in decline and will fall sharply in urban areas as new automotive technology takes over.

Urban air quality is improving rapidly, and clean air in our cities will be a reality in the next few years.

BIBLIOGRAPHY

Calvert, J.G., Heywood, J.B., Sawyer, R.F., Seinfeld, J.H., 1993. Achieving acceptable air quality: Some reflections on controlling vehicle emissions. *Science* 261, 37–45.

International Council on Clean Transportation, 2019. *Gasoline versus diesel: Comparing CO_2 emissions levels of a modern medium size car model under laboratory and on-road testing conditions.* Washington, DC: International Council on Clean Transportation.

International Gas Union, 2018. *Urban air quality in Chinese and South American cities dramatically improved by switching to natural gas.* Barcelona, Spain: International Gas Union.

Nagl, C., Ansorge, L., Moosmann, L., Spangl, W., Wiesenberger, H., 2013. Critical areas for compliance with PM_{10} and NO_2 limit values. In M. Viana (ed.), *Urban air quality in Europe*, pp. 3–30. Heidelberg, Germany: Springer.

Pearson, J.K., 2019. European solvent VOC emissions inventories based on industry-wide information. *Atmospheric Environment* 204, 118–124.

Samaras, Z., Zafiris, D., Pethainos, D., Zierock, K.-H., 1993. Forecast of road traffic emissions in the European Community up to the year 2000. *Science of the Total Environment* 134, 251–262.

Simpson, D., *et al.*, 1999. Inventorying emissions from nature in Europe. *Journal of Geophysical Research* 104, 8113–8152.

World Health Organization, 2021. *Air quality guidelines: Global update 2021.* Copenhagen, Denmark: Regional Office for Europe of the World Health Organization.

5

REGIONAL AIR QUALITY

All the best science in the world, without translation into policy, really is of no practical value in the world of tomorrow.
– Professor Sir Leszek Borysiewicz, vice-chancellor of the University of Cambridge, October 2010–October 2017

OVERVIEW

In this chapter, we explore how concerns about regional air quality began with the discovery of the effects of acid rain, where environmental damage in one country was due to air pollution transported from other countries. These concerns led to the development of regional air quality models. The important regional air pollutant, ozone, is then examined and how its reduction is necessary to improve urban and regional air quality and ease the build-up of global warming.

Emissions inventories are examined and how they can be used to assist environmental bodies and governments by providing input data for air quality models and formulate policies to improve future air quality. The role of air quality legislation is explored and how international bodies can work together to reduce regional air pollution.

The environmental issues that come under the heading of regional air pollution include acid rain, ground-level ozone and fine particle formation, eutrophication of sensitive ecosystems and the loss of biological diversity, and the long-range transport of heavy metals (HMs) and persistent organic pollutants (POPs). Each of these issues involves environmental impacts at often large distances,

DOI: 10.4324/9781003293132-5

perhaps over hundreds to several thousand kilometres, downwind of major sources of pollutants. Because of the huge distances involved, travel times may be days to weeks. Inevitably, important pollutant emission sources may be in one country, state or jurisdiction and the environmental impacts in another. In many cases, the benefits of environmental improvement may be realised in one country whereas the costs of pollution controls are borne in another.

We begin with the acid rain problem. Svante Oden, a Swedish scientist in 1967 described a new environmental problem: acid rain. He pointed to the increasing acidity of rainwater and of the lakes and streams in southern Sweden and the associated near-extinction of fish in these sensitive surface waters. Sulphur dioxide (SO_2) emissions from the large power stations in the United Kingdom, Germany and the USSR turned out to be the source of the acidic sulphur species that were driving the acidification problems in southern Sweden. Long-range transport was eventually shown to be the cause of these acidification problems, together with those identified later in southern Norway, Finland, Dumfries and Galloway in southern Scotland, and Snowdonia in Wales.

The first regional air pollution model, the EMEP model, was built by a Norwegian scientist, Anton Eliassen, in 1977, and was used to construct 'blame matrices' which allowed the deposition of acidic sulphur species in sensitive Scandinavian environments to be attributed to the SO_2 emissions from each European country. These 'blame matrices' were then built into an integrated assessment model (RAINS) by Joseph Alcamo of the International Institute for Applied Systems Analysis (IIASA) in Vienna in 1983 by extending the regional SO_2 model in two distinct directions. The treatment of SO_2 emissions was extended to address information of the costs of pollution control measures and the treatment of deposition was extended to address the critical loads of the receiving ecosystems. The critical load of an ecosystem was defined as the maximum tolerable deposition load that the receptor could withstand without damage occurring. The integrated assessment models thus allowed critical loads to be reached while sharing the costs of controls across the major polluting countries.

Acid rain damage has been found in eastern North America in Ontario, Canada and in the north-east of the USA, being first discovered by Gene Likens at Hubbard Brook in the White

Mountains of New Hampshire in 1963. Lake acidification and the loss of fish stocks have been sufficiently well documented that acid rain became an important policy issue for both US and Canadian administrations. The transboundary nature of the problem was recognised quite early, with a particular focus on the large coal-burning power plants, steel mills and petrochemical plants in the Ohio valley. The USA adopted a characteristically different policy approach to acid rain control from that adopted in Europe, preferring one based on emissions trading, rather than integrated assessment modelling. Large electricity-generating power stations purchased emission permits with the price of the permits increasing with time, forcing in SO_2 emission abatement measures in a cost-effective manner.

Ground-level O_3 formation has been a long-standing air pollution problem in the USA that was initially considered an urban problem rather than a regional one. Initially, attention focussed on the Los Angeles basin but then elevated O_3 levels were found to occur in all major population centres in the contiguous USA. Consequently, a state-wide approach was followed rather than a federal approach, with California very much in the lead both from science and policy perspectives. For example, John Seinfeld from the California Institute of Technology in 1973 put together the first set of VOC and NO_x emission inventories, together with the first complete chemical mechanism for photochemical smog formation to assemble a three-dimensional grid model of O_3 across the Los Angeles basin. This study was a forerunner of urban air quality models. States formulated their own strategies for tackling elevated O_3 levels and for managing their exceedances of the US National Ambient Air Quality Standard for O_3. Following the Clean Air Act and its amendments in the 1970s, the federal government took increased responsibility for the air quality goals and standards to be set and implemented nationwide. After three decades of action at the federal and state levels, 68 areas were still in non-attainment by 1987. Long-range, interstate O_3 transport emerged during the 1980s as a particular issue for O_3 non-attainment in the north east US. Currently, 50 urban areas with a total population of 122 million were classified as breaching the US National 8-hour Air Quality Standard. In contrast, only six urban areas representing 20 million people were in breach of the US national $PM_{2.5}$ standard.

The German scientist Bernard Ulrich raised the problem of forest dieback in 1980 and shifted the focus of the acid rain debate in Europe onto ground-level O_3. Elevated ozone levels had been detected during sunny, summer weather across Europe from the 1970s onwards. James Lovelock demonstrated how, at a location on Bantry Bay in the remote south-west of Ireland, elevated levels of haze and ground-level ozone were associated with elevated levels of CFC-11, a unique man-made halocarbon. In this way, Lovelock showed how ground-level ozone could travel hundreds to thousands of kilometres, crossing national and international borders. Episodes of elevated O_3 levels were increasingly associated with human health effects as evidenced by the exceedance of internationally accepted air quality standards, together with damage to crops and vegetation and with forest dieback. Ground-level O_3 formation occurs when sunlight-driven chemical reactions in the atmosphere transform the hundreds to thousands of organic compounds into aggressive secondary pollutants such as O_3 and PAN, in the presence of NO_x. Again, these atmospheric chemical reactions take hours and days to complete and so ground-level O_3 formation is seen as an archetypal regional and trans-boundary air pollution problem.

The acid rain problem in Europe had brought the European countries to the negotiating table and ushered in scientific research cooperation, regional air pollution modelling and integrated assessment modelling under the aegis of the United Nations Economic Commission for Europe (UNECE) in Geneva, Switzerland. However, the policy pressure to address simultaneously SO_2, NO_x and VOCs emissions and to focus simultaneously on both acidification and ground-level ozone formation, meant that the single pollutant approach was no longer viable and would be too cumbersome to address future concerns about fine particle formation and eutrophication that were on the horizon. The Executive Body of the Geneva Convention on Long-Range Transboundary Air Pollution (CLRTAP) came up with the so-called multi-pollutant multi-effect approach to address the growing number of regional pollution problems faced by Europe in the 1990s onwards. The policy framework underpinning this multi-pollutant multi-effect approach is illustrated in Figure 5.1.

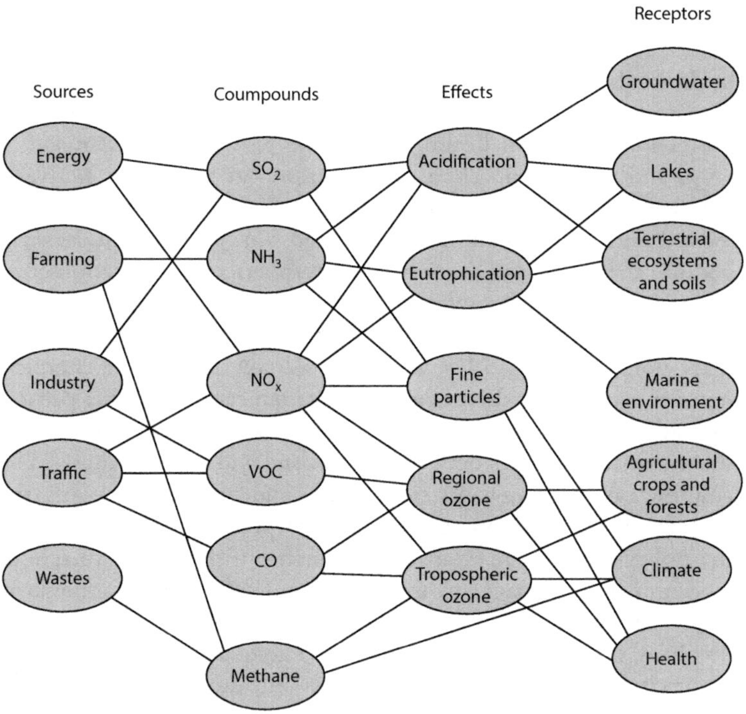

Figure 5.1 The multi-pollutant multi-effect framework underpinning our understanding of regional air pollution problems.
Source: redrawn from Grennfelt et al. (1994), by permission of Springer Nature.

A European policy-maker would begin at the right-hand side of the framework diagram with the environmental damage column. Policy-makers express commitments to reducing damage in these receptor systems within their own jurisdictions. Environmental scientists identify the causative agents that are causing the damage and these are listed in the third column of the framework, with lines showing the linkages between the causative agents and the environmental damage. Environmental acidification links to lakes, groundwaters and soils because this is the mechanism driving the loss of fishery status. The over-supply of nutrients (eutrophication) links to the loss of biological diversity in soils and plant ecosystems and to the formation of algal blooms in shallow seas such as the Baltic Sea and in lakes. Fine particles ($PM_{2.5}$) damage human health and contribute to global climate change. Elevated regional O_3 levels lead to exceedance of air quality standards set for the protection of human health and for the prevention of damage to crops and vegetation. There is also an important synergy between regional O_3 and global climate change. The linkages between the second and third columns are based on atmospheric science because these linkages record the atmospheric transformations that convert primary pollutants in the second column to the causative agents or secondary pollutants in the third column. Thus, the primary pollutants SO_2, NO_x and NH_3 all take part in the chemical transformations that drive acid rain, eutrophication and fine particle formation. VOCs and NO_x drive ground-level ozone formation as shown by the linkages between the second and third columns. The column on the left-hand side of the framework represents the sources of the emissions of the primary pollutants. This is where the policy-makers can exert their influence if they want to reduce environmental damage on the right-hand side of the framework. The framework identifies a wide range of emission sources involved with regional air pollution. These are not just the large electricity generating power stations but motor vehicle traffic, agriculture and industry.

This policy framework in Figure 5.1 was then implemented in the regional EMEP air pollution model and the RAINS integrated assessment model used in the negotiations underpinning the multi-pollutant multi-effect Gothenburg Protocol to the UNECE CLRTAP Convention in 1999. It was soon realised in these negotiations that just doing what is planned by countries is not enough but achieving the planned reductions is better for the regional environment than doing nothing. Achieving the maximum feasible reductions would be worthwhile but it would be expensive. For all the environmental damage categories in the above framework, it is the case that a cooperative approach saves effort and money. In the twenty years since the Gothenburg Protocol much has been achieved across Europe. Initially the multi-pollutant multi-effect approach focussed on ecosystem effects but in 2012 health effects were added through the additional consideration of fine particles. Since then, human health effects from fine particles have largely taken over from ecosystem effects as the main driver for the consideration of strategies aimed at the improvement of regional air quality across Europe.

As a result of the UNECE CLRTAP Convention, much has been achieved in terms of reducing SO_2 emissions and thereby the deposition of acidic substances. This has yet to feed through into improvements in soil and freshwater acidification and species recovery because the acidification effects of historical emissions will remain for decades. Progress in terms of reducing NO_x emissions has been much slower and there has been little change in agricultural NH_3 emissions. Eutrophication and loss of biological diversity therefore remain significant problems. Fitting exhaust gas three-way catalysts and evaporative canisters to gasoline motor vehicles has dramatically reduced VOC emissions. Episodic peak summer-time ozone levels have therefore fallen dramatically in terms of the number and severity of regional ozone pollution episodes, see Figure 5.2. This figure shows how the peak hourly mean O_3 levels monitored in the UK and across Europe have declined over five decades. However, despite the reduction in peak levels,

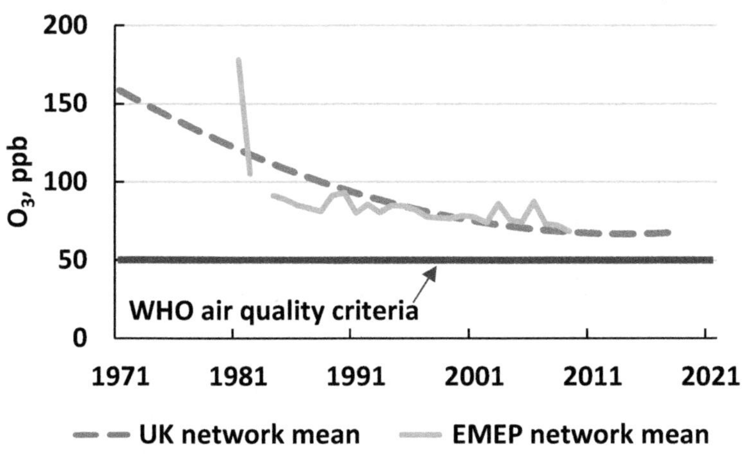

Figure 5.2 Maximum 8-hourly mean ozone concentration averages for 11 rural UK and 57 long-running EMEP rural ozone monitoring stations from 1971 to 2019, together with the WHO ozone air quality criteria value set for maximum 8-hour mean ozone exposures.

exceedances of the 2021 World Health Organization (WHO) ozone air quality criteria still persist.

Over Europe as a whole, 15,100 premature deaths have been attributed to elevated O_3 levels. Premature deaths are highest in Italy, Germany, Spain and France and on a per capita basis are highest in Greece, Albania, Kosovo, Italy and Montenegro. Internationally agreed air quality standards for O_3 set to protect human health are still exceeded across Europe because the ambition levels agreed for the reduction of NO_x emissions from road traffic have not been set high enough.

More recently, it has been realised that for a full treatment of the regional O_3 issue at some point in the future, a larger geographic area than Europe alone will be required. This is because there is an important contribution to the O_3 levels in Europe from hemispheric scale long-range air pollution transport. Future policies to control O_3 in Europe will need to take the hemispheric scale into account and address the contributions of elevated O_3 levels in Europe from North America and Asia. It is not clear how such inter-regional discussions could be formulated within the current international environmental frameworks. We will return to these challenges in Chapter 7.

During the negotiations underpinning the UNECE CLRTAP Gothenburg Protocol, the issue of eutrophication (the oversupply of nutrients) has surfaced and has become an increasingly important policy concern. As the linkages in Figure 5.1 demonstrate, eutrophication is driven by the deposition of ammonia, ammonium sulphate and ammonium nitrate, so-called reduced nitrogen, on sensitive environments such as Gelderland in the Netherlands. Some plant species respond well to the increased deposition loads of reduced nitrogen but others cannot cope and are driven out of sensitive ecosystems. This leads to the loss of biological diversity and the disappearance of the so-called 'red list' plants. It has proved difficult to reduce ammonia emissions from agriculture across Europe because of the competing policies implied by the EU National Emissions Ceilings Directives and the EU Common Agriculture Policy. Eutrophication and agriculture emissions have become intractable problems for regional air quality in Europe.

EMISSION INVENTORIES

Just as we have a detailed emissions inventory for vehicles in urban regions that is revised on a regular basis, we require a detailed emissions inventory for each region as illustrated in the left-hand column of Figure 5.1. Regional emission inventories are typically established on detailed 10 km × 10 km grids covering the region in question and all major sources of pollution are included. Separate inventories are developed for each pollutant: SO_2, NO_x, VOCs, NH_3, PM, CO, CH_4, N_2O and CO_2.

Typically, the inventory would include the following sector categories:

1 public power, co-generation, and district heating plants;
2 commercial, institutional, and residential combustion plants;
3 industrial combustion plants and processes;
4 non-combustion processes;
5 extraction and distribution of fossil fuels;
6 solvent use;
7 road transport;
8 other transport;
9 waste treatment and disposal;
10 agriculture; and
11 nature (biogenic).

Point source emissions will also be included, such as:

1 combustion plants and refineries;
2 sulphuric and nitric acid plants;
3 paper pulp production plants;
4 workshops and steel plants;
5 car painting plants;
6 international airports; and
7 any combustion plants with chimneys heights over 100 m.

Often, man-made emissions from point sources are calculated from the equation:

Emission = emissions factor × activity level × control factor.

For power station emissions, the emissions factors are related to the mass of fuel burned, the activity level would be the fraction of annual use, and the control factor would relate to any improvement in pollution control. Both power stations and refineries are changing rapidly – replacing a coal-fired power station with natural gas halves the CO_2 emissions factor. Refineries are being closed worldwide as oil demand declines in the 2020s. Industry has a key role in updating their emissions inventories.

For VOC emissions, the solvents business has been proactive in eliminating species which impair air quality and also demonstrated that legislative bodies overestimate solvent VOCs, particularly in the European Union.

It is essential to continue to update emissions inventories to guide legislative authorities and provide accurate emissions data for use in the mathematical models used for regional air quality policy and strategy development based on Figure 5.1. Continuous improvement of emissions inventories ensures that regional air quality meets safer threshold limits of air pollution and leads to improved human health. Appendix 3 describes in more detail the principles used to build regional air quality models addressing country budgets for acidic sulphur and oxidised nitrogen, together with ground-level ozone formation.

HEALTH AND LEGISLATION

The principal authority for providing guidelines on threshold limits for key air pollutants that cause health risks is the World Health Organization (WHO). In 2021, WHO published a list of exposure levels which should not be exceeded for $PM_{2.5}$, PM_{10}, O_3, NO_2 and SO_2. These provide a reference for setting external air pollution targets at regional and national levels, and are shown in Table 5.1.

Each country can adopt its own approach to setting regional and urban air quality standards and, necessarily, the air quality targets differ. In the United States of America, the Environmental Protection Agency (EPA) has the responsibility for setting and issuing air quality limits for each pollutant. The European Commission also sets air quality limit values and these are agreed by all 27 member states, a process which involves intensive consultation with industry, air quality experts and individual governments.

It is interesting to compare the different approaches to legislation of pollutants adopted by the United States and the European

Table 5.1 WHO (2021) guidelines on exposure limits for the main air pollutants.

Species	*1-hour mean*	*8-hour mean*	*24-hour mean*	*Annual mean*
SO_2	500 μg m^{-3}*		20 μg m^{-3}	
NO_2	200 μg m^{-3}		25 μg m^{-3}	10 μg m^{-3}
O_3		100 μg m^{-3}		
$PM_{2.5}$			15 μg m^{-3}	5 μg m^{-3}
PM_{10}			45 μg m^{-3}	15 μg m^{-3}

*10-minute mean

Union, particularly with respect to O_3. The European Union approach aims to control and reduce the emissions of all VOCs, whereas the United States is concerned about the emissions of the most highly reactive of the VOC species. The difference arises because historically, the USA situation has been dominated by concerns over the intense urban-scale photochemical ozone production seen in the Los Angeles basin and in Houston, Texas. In these situations, only the most highly reactive of the VOC species contribute. In contrast in Europe, ground-level O_3 has always been seen as a transboundary phenomenon and, in principle, VOCs of different reactivity may contribute, to a greater or lesser extent.

The European approach is laid out in some detail in the 1991 Geneva VOC Protocol to the Convention on Long-Range Transboundary Air Pollution. The VOC Protocol recognises that VOCs differ greatly from each other in their reactivity and countries should focus their efforts on reducing emissions of those VOC emissions that contribute most to ground-level O_3 formation. The Protocol goes on to describe a classification of VOCs into three groups according to their importance in episodic O_3 formation and this classification is presented in Table 5.2.

We can use the VOC classification in Table 5.2 to link together the European and USA approaches. The VOC species classified as 'more important' include the alkenes: ethylene and 1,3-butadiene, which are the VOCs that drive the intense urban-scale photochemical ozone production in Houston, Texas. Also, the alkenes and aromatics have important motor vehicle sources that drive

Table 5.2 Classification of VOCs into three groups according to their importance in episodic ozone formation as laid out in the 1991 Geneva VOC Protocol to CLRTAP.

More important	
alkenes	
aromatics	such as xylene
alkanes	$>C_6$ alkanes except 2,3-dimethylpentane
aldehydes	all aldehydes except benzaldehyde
biogenics	isoprene
Less important	
alkanes	$C_3 - C_5$ alkanes and 2,3-dimethylpentane
ketones	MEK and methyl-t-butyl ketone
alcohols	ethanol
esters	all esters except methyl acetate
Least important	
alkanes	methane and ethane
alkynes	acetylene
aromatics	benzene
aldehydes	benzaldehyde
ketones	acetone
alcohols	methyl alcohol
esters	methyl acetate
chlorinated hydrocarbons	methyl chloroform, methylene chloride, trichloroethylene, tetrachloroethylene

urban-scale ozone formation in the Los Angeles basin. This classification also explains why unreactive or 'least important' VOCs such as acetone have been deregulated by the US EPA.

Biogenic sources of VOCs such as isoprene, also play an important role in ground-level O_3 formation and hence isoprene appears in the 'more important' class in Table 5.2. Xylene is a highly reactive VOC species which comes within the aromatics class. It is an important component of some widely used solvents and hence it has been the focus of attention within the solvents industry.

We will examine ozone in further detail when we discuss the challenges surrounding global air quality and climate change in Chapter 6.

BIBLIOGRAPHY

Carter, W.P.L., 1994. Development of ozone reactivity scales for volatile organic compounds. *Journal of the Air and Waste Management Association* 44, 881–899.

Climate & Clean Air Coalition, 2019. Air pollution in Asia and the Pacific: Science-based solutions (summary + full report). https//ccacoalition.org/en/resources/air-pollution-asia-and-pacific-science-based-solutions-summary-full-report.

Cox, R.A., Eggleton, A.E.J., Derwent, R.G., Lovelock, J. E., Pack, D.H., 1975. Long-range transport of photochemical ozone in north-western Europe. *Nature* 255, 118–121.

Derwent, R.G., Jenkin, M.E., Saunders, S.M., Pilling, M.J., 1998. Photochemical ozone creation potentials for organic compounds in north west Europe calculated with a master chemical mechanism. *Atmospheric Environment* 32, 2429–2441.

Fiore, A.M., *et al.*, 2012. Global air quality and climate, *Chemical Society Reviews* 41, 6663–6683.

Grennfelt, P., Hov, O., Derwent, R., 1994. Second generation abatement strategies for NO_x, NH_3, SO_2 and VOCs. *Ambio* 23, 425–433.

Lovelock, J.E., 1972. Atmospheric turbidity and CCl_3F concentrations in rural southern England and southern Ireland. *Atmospheric Environment* 6, 917–925.

Monks, P.S., *et al.*, 2015. Tropospheric ozone and its precursors from the urban to the global scale from air quality to short-lived climate forcer. *Atmospheric Chemistry and Physics* 15, 8889–8973.

National Research Council, 1991. *Rethinking the Ozone Problem in Urban and Reginal Air Pollution*. Washington, DC: National Research Council Academic Press.

Wuster, H., 1992. The Convention on Long-range Transboundary Air Pollution: Its achievements and its potential. In T. Schneider (ed.), *Acidification research: Evaluation and policy applications*. Amsterdam, The Netherlands: Elsevier Science Publishers.

6

GLOBAL AIR QUALITY AND CLIMATE CHANGE

OVERVIEW

This chapter identifies the key pollutants that drive climate change and, importantly, the timescales involved. The three most important man-made greenhouse gases are identified as carbon dioxide, methane and ozone with atmospheric lifetimes of over 100 years, 12 years and 20 days respectively. We predict that since pre-industrial times, these same three gases have raised global warming by 1.04° C, 0.34°C and 0.21°C respectively by 2020. We also demonstrate that aerosols, notably ammonium sulphate and nitrate have contributed to a global cooling of 0.46°C over the same period. Other pollutants are also examined to give a complete overview of global air quality and potential climate change. These results suggest that we have left it too late to meet the global temperature target set by the Paris Agreement of 1.5°C and while air quality will unquestionably improve in the future, climate change will remain a considerable challenge.

There is no question that global climate change is a complex phenomenon that is of undoubted importance to all mankind. It presents a number of challenges, not only in scientific understanding, but also in the magnitudes of the changes that will be required to life as we know it, if global climate change is to be brought under control and mitigated. In this chapter, we identify the key processes that drive climate change and come to grips

DOI: 10.4324/9781003293132-6

with the timescales involved. We begin by describing how global air pollutants interact with the global climate system.

Since the industrial revolution, mankind has changed the composition of the atmosphere through emissions of certain key air pollutants. These pollutants, the greenhouse gases and aerosols have intercepted the outgoing infrared radiation leaving Earth and disturbed Earth's radiation balance with the incoming radiation from the sun. The climate system has responded to this out-of-balance by warming Earth's surface and atmosphere and so regaining the radiation balance. This warming is the so-called greenhouse effect and it is said to have been forced by the atmospheric composition change. This forcing mechanism is called radiative forcing and the Earth–atmosphere system response is climate change.

The key processes involved with global climate change are illustrated in the framework diagram shown in Figure 6.1. The arrows pointing downwards represent cause and effect linkages and illustrate the various forcing mechanisms. Emissions of greenhouse gases and

EMISSIONS FROM HUMAN ACTIVITIES

⇓

ATMOSPHERIC CONCENTRATION CHANGES

⇓

RADIATIVE FORCING

⇑⇓

PHYSICAL RESPONSE

⇓

CLIMATE RESPONSE

⇓

BIOLOGICAL RESPONSE

Figure 6.1 Framework diagram representing the key processes involved with climate change.

aerosols from human activities cause changes in the composition of the atmosphere. These atmospheric composition changes since pre-industrial times, interfere with Earth's radiation balance and lead to radiative forcing. Earth's system responds to this forcing by warming and this is the physical response shown in the framework diagram above. A warmer atmosphere holds more water vapour, may have less polar ice, may be stormier and have more extreme weather events. All these responses when taken together comprise a climate response to the radiative forcing and to the atmospheric composition change. Biological systems respond to this climate response because they find their growing conditions have changed, becoming more or less favourable and this is the biological response at the bottom of the framework diagram.

While we are absolutely sure of the atmospheric composition changes since pre-industrial times, through measurements of the greenhouse gases in ice cores, uncertainties increase quickly as we move down the framework diagram. This is to be expected as we move from aspects of physics and chemistry through to those of biological and ecological sciences. There is an inordinate amount of complexity in the global climate system and we are a long way off a complete description of how the global climate system will respond to the atmospheric composition change. Yet, we can be sure that it will respond.

We have assembled a global climate model to describe the elements of Figure 6.1 so that we can develop some understanding of the challenges surrounding global climate change. The details of our global climate model are presented in Appendix 4 and how it can be used to address carbon dioxide (CO_2), methane (CH_4), nitrous oxide (N_2O), tropospheric ozone (O_3), halocarbons and aerosol particles.

We have assembled some simple tools to describe the global climate system and we use them to examine the greenhouse gases and aerosols and their roles in driving global climate change. In this way we can begin to understand how firm our understanding is of radiative forcing and global climate change and the timescales involved. Three factors exert a major influence on the relative roles played by man-made air pollutants in global climate change. The first factor is the magnitude of man-made emissions. The second factor is the residence time of the pollutant in the atmosphere. The third factor is radiative efficiency, the ability of the pollutant to absorb the out-going infrared radiation from Earth's surface–atmosphere system. CO_2 is an important man-made greenhouse gas because its emissions are huge and it has a long atmospheric lifetime of hundreds to thousands of years. Despite having a much shorter lifetime of about 12 years, CH_4 is an important man-made greenhouse gas because it also has huge emissions and has a high radiative efficiency. N_2O has relatively small man-made emissions compared with CO_2 but it has a long atmospheric lifetime of about 120 years and like CH_4 has a high radiative efficiency. O_3 has an extremely short atmospheric lifetime of 20 days but is an important man-made greenhouse gas because of its huge radiative efficiency and an enormously large atmospheric turnover. Halocarbons are important greenhouse gases because they are strong absorbers in the infrared with long atmospheric lifetimes but, fortunately, relatively small man-made emissions. Aerosols lead to negative radiative forcing and hence they cool the global climate system. Their lifetimes are short, up to 10 days or so, turnovers are moderate but they exert a huge cooling influence on the climate system by scattering the incoming solar radiation.

We make a start with carbon dioxide (CO_2). In many ways, CO_2 is the most important man-made greenhouse gas and it is certainly the most complex. However, aspects of its fate and behaviour are well-understood and so it makes a good place to start. Then we move on to CH_4, N_2O, O_3, halocarbons and the aerosols.

ATMOSPHERIC CARBON DIOXIDE

CO_2 emissions arise largely from the combustion of fuels. Some fuels, the so-called fossil fuels, such as coal, oil and natural gas contain carbon that was laid down in geological times. Its combustion therefore adds 'new' CO_2 into the atmosphere and therefore contributes directly to atmospheric composition change and hence global climate change. Other fuels such as wood, peat, animal manure, crop wastes and biogas contain carbon that has been recently derived from living plants and trees. Its combustion adds no 'new' CO_2 to the atmosphere because the plants and trees that acted as the source of the carbon, had produced this carbon by photosynthesis, taking up CO_2 from the atmosphere. Combustion of fuels derived from biomass contribute no 'new' CO_2 to the atmosphere and are climate-neutral, in contract to fossil fuel combustion which does contribute 'new' CO_2 and is climate-warming.

Because of the long lifetime of CO_2, man-made emissions of CO_2 remain in the atmosphere for extended periods. For every ppm of CO_2 emitted into the atmosphere in one year, 72% remains airborne after 10 years, 40% after 100 years and 10% after 1000 years. As a consequence, it is important to think in terms of the cumulative emissions of man-made CO_2.

At the start of the industrial revolution in 1750, the cumulative emissions of man-made CO_2 amounted to 9.4 million tonnes. Almost all of this was from sea coal combustion in the UK. By 1820, the cumulative emission had reached 1.58 billion tonnes, all of this from coal combustion, of which 94% occurred in the UK. As shown in Figure 6.2, there were also small contributions from Germany, France and the USA. By 1920, cumulative emissions had reached 103 billion tonnes, with the USA as the largest country then, with a 35% share and the UK in second place with 25%. Cumulative emissions from the USA had overtaken the UK in 1912. Again, coal combustion was by far the largest source, accounting for 90% of man-made CO_2 emissions in 1920. By 2020, cumulative emissions had reached 1650 billion tonnes. The USA was still the largest cumulative source but its share had been reduced to 25%. China, Russia and Germany were second, third and fourth most important cumulative emitters as shown in Figure 6.2. Germany overtook the UK in 1981 and China in 1999. Even in 2020, coal still accounts for the majority of cumulative emissions, with petroleum in second place and natural gas in third.

In 2020, almost every country worldwide reported some man-made CO_2 emissions because of the widespread use of petroleum-based fuels for transport. Indeed, of the 2020 man-made CO_2 emission of 36 billion tonnes worldwide, 12.36 billion tonnes came from petroleum-based fuels, 14.35 billion tonnes from burning coal in large electricity-generating power stations, 7.62 billion tonnes from burning natural gas and 1.56 billion tonnes from cement manufacture.

Armed with these historic CO_2 emission estimates and using our global climate model, we can put together a picture of the time history of atmospheric CO_2 concentrations. With some estimates as to future CO_2 emissions beyond 2020, it is straightforward to

Figure 6.2 Pie charts showing cumulative historic CO_2 emissions by country in 1820, 1920 and 2020.

plot the trajectory of future atmospheric CO_2 concentrations. Later, we will use the available emissions projections to test mitigation options and make some sensible projections. At this stage, we are interested in using the simple model to illustrate the behaviour of each greenhouse gas and so it is sensible to use a more idealistic future scenario that hopefully will be more revealing. We therefore chose a future beyond 2020 with absolutely no CO_2 emissions so that we can demonstrate the long timescales involved with this greenhouse gas. Figure 6.3 shows the time history of model atmospheric CO_2 concentrations over the period from 1950 to 2050.

Our global climate model slightly under-predicts the atmospheric CO_2 level in 1950 by 15 ppm at 313 ppm or about 5% which is well within the errors of the historic CO_2 emissions. By 2014, the model underestimation is considerably reduced to within 2 ppm at 398 ppm. Considering the accuracy of current and historic CO_2 emissions, we take the performance of our global climate model for CO_2 to be satisfactory, for illustrative purposes. On this basis, Figure 6.3 gives a representation of future CO_2 levels beyond 2020, making the assumption that there were absolutely no further CO_2 emissions. We see that the model curve shows an abrupt change in slope after 2020 reflecting the cessation of atmospheric emissions. CO_2 levels fall after 2020 but the rate of fall is quite slow. Indeed by 2050, after 30 years of zero emissions, CO_2 levels have only fallen by 30 ppm, that is by only 7%. It is the long lifetime of CO_2 that

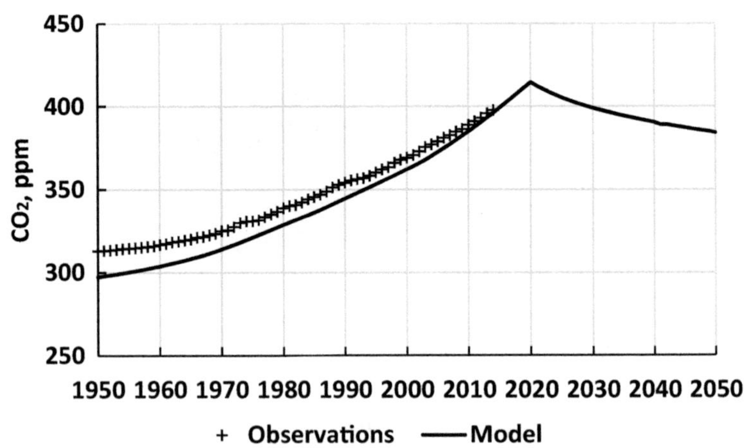

Figure 6.3 The observed and model predicted atmospheric CO_2 levels from 1950–2050.

hinders the response of CO_2 levels to a cessation of emissions and hence makes its contribution to radiative forcing and global climate change so difficult to mitigate by emission reductions. A further testament to the long lifetime of CO_2 is that, of the 1650 billion tonnes of cumulative emissions since pre-industrial times, just close to two-thirds still remain airborne in 2020.

ATMOSPHERIC METHANE

CH_4 is a potent greenhouse gas because of a combination of large global man-made emissions and a high radiative efficiency. The most important man-made source of CH_4 is from agriculture through emissions from rice paddies and from livestock. Both these sources involve bacterial production either in water-logged soils, in animal digestive systems or from animal waste handling and storage. Fossil fuel production and use is the second major CH_4 source and includes emissions from coal mining and from the extraction and use of natural gas and oil. CH_4 is the major constituent of natural gas and so natural gas leakage from its transmission and distribution networks is an important source of atmospheric CH_4. Sources count as CH_4 sources when they lead to atmospheric CH_4 release but as CO_2 sources when burnt and lead to CO_2 release, so as not to cause double-counting.

CH_4 has some clear-cut natural sources which include tropical wetlands, bogs, marshes, oceans, vegetation, wild animals, termites, geological leakage and permafrost melting. Climate change may exert an influence on these emissions. Indeed, man-made climate change induced permafrost warming could be catastrophic and lead to a runaway greenhouse effect. Some CH_4 sources have been difficult to categorise as either human-influenced or natural and this particularly applies to biomass burning. Here, incomplete combustion can lead to CH_4 release. Man-made sources currently account for 57% of CH_4 emissions and natural sources about 38%, leaving about 5% as undecided, out of a total of about 600 Tg yr^{-1}.

In Figure 6.4, the historic emissions of CH_4 have been fed into our global climate model, making the assumption that emissions cease after 2020. The conceptual model is well able to account accurately for the growth in CH_4 concentrations since pre-industrial times. After 2020, CH_4 concentrations fall rapidly in response to the

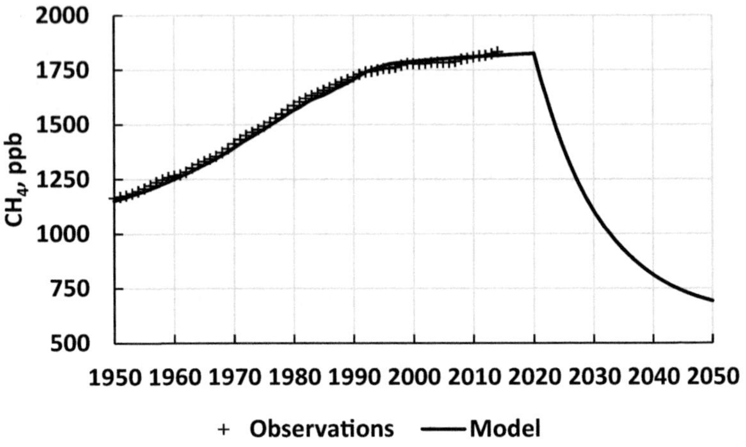

Figure 6.4 The observed and model predicted global atmospheric CH_4 levels from 1950–2050.

idealised scenario because of the short, 12 years, atmospheric lifetime of this greenhouse gas. This means that, in principle, CH_4 concentrations will respond quickly to future mitigation options.

ATMOSPHERIC NITROUS OXIDE

Nitrous oxide (N_2O) is an important greenhouse gas by virtue of its long atmospheric lifetime, 120 years, and its high radiative efficiency. N_2O has appreciable natural sources which maintained a steady atmospheric level of 274 ppb since the last ice age and up to pre-industrial times. These natural sources amount to about 41 Tg N_2O yr^{-1} and involve bacterial processes in upland soils, banks of rivers, estuaries, coastal water and oceans. Since the industrial revolution, man-made sources have driven up N_2O emissions dramatically. Soils under intensive agriculture give rise to between 14 and 17 Tg N_2O yr^{-1} through the use of organic (animal wastes) and synthetic ammonia- or urea-based fertilisers, representing two-thirds to three quarters of all man-made sources.

The remaining man-made N_2O sources are the industrial processes producing nitric acid for the manufacture of ammonia- and urea-

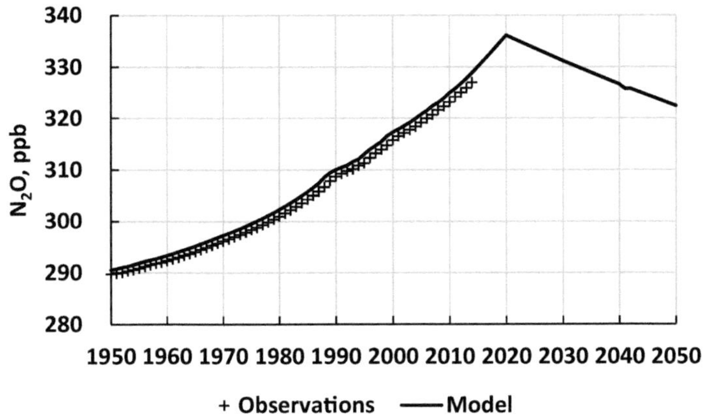

Figure 6.5 The observed and model predicted atmospheric N_2O levels from 1950–2050.

based fertilisers and adipic acid for the manufacture of synthetic fibres such as nylon. Biomass burning and fossil fuel combustion are additional sources of N_2O. Together these industrial sources account for between 4 and 7 Tg N_2O yr^{-1}.

Figure 6.5 presents the observed and model predicted atmospheric levels of N_2O from 1950 through to 2050 using our understanding of its sources and sinks. The agreement between our global climate model predictions and the observations is good. The assumption has been made that man-made N_2O emissions have been completely curtailed after 2020 to illustrate the importance of its long atmospheric lifetime. Levels decline by only 4% between 2020 and 2050 in response to a cessation in emissions and this idealistic scenario illustrates the relative inertia inherent in N_2O responses to mitigation measures.

HALOCARBONS

Halocarbons are a group of up to forty organic compounds containing fluorine, chlorine, bromine or iodine that are widely used as aerosol propellants, refrigerants, air conditioning fluids, foam-blowing agents, anaesthetic gases and solvents. Each has its own emission

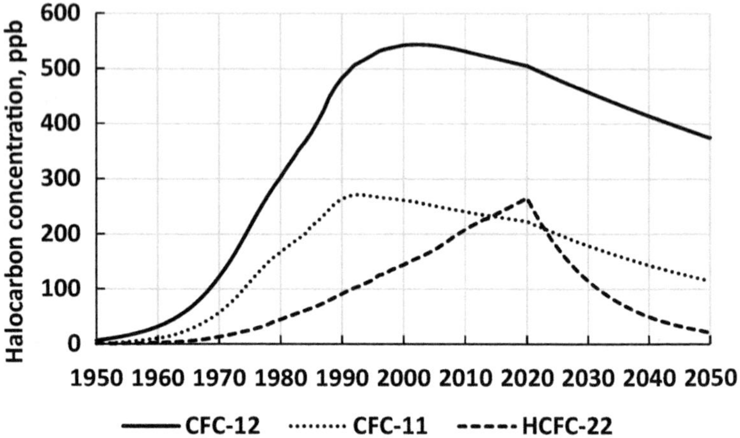

Figure 6.6 The model predicted atmospheric levels of three selected halocarbons from 1950–2050.

pattern, atmospheric lifetime and radiative efficiency and hence ability to act as a greenhouse gas. Almost all, with a few notable exceptions, are entirely synthetic and are not present in nature and have only been present in the atmosphere in detectable levels since 1950.

Figure 6.6 presents our global climate model predictions for three of the most important halocarbons: CFC-11, CFC-12 and HCFC-22, as an illustration of this class of compounds. CFC-11 and CFC-12 were used as aerosol propellants and refrigerants until their phase-out under the Montreal Protocol. Their levels are slowly and steadily declining under the influence of their long atmospheric lifetimes of 100 and 45 years, respectively. HCFC-22 is a widely used air conditioning fluid, mainly in through-the-wall units. Its atmospheric lifetime is 12 years as shown by its rapid decline in Figure 6.6, following the assumption of a complete curtailment of emissions in 2020.

TROPOSPHERIC OZONE

Ozone (O_3) has both natural and man-made sources and is widely distributed throughout the upper and lower atmospheres. In the upper atmosphere or stratosphere, O_3 absorbs the harmful ultraviolet radiation from the sun and provides a sun-screen that protects all life on Earth. Halocarbons have damaged this O_3 layer as mentioned in Chapter 2. The O_3 in the stratosphere, however, is not a major issue for global climate change and we will leave this issue for the present and turn our attention to O_3 in the lower atmosphere.

In the lower atmosphere or troposphere, O_3 has both natural and man-made sources. O_3 is a secondary pollutant and all the O_3 present there has been formed by sunlight-driven photochemical reactions. These atmospheric chemical reactions are an absolutely phenomenal source of ozone and turnover the whole O_3 burden of the troposphere in twenty days or so. They represent the largest turnover of any pollutant, being of the order of 10 million tonnes per day. These photochemical reactions involve a number of ozone precursors: CH_4, carbon monoxide (CO), hydrogen (H_2), myriads of organic compounds and the oxides of nitrogen (NO_x). Because of the importance of manmade sources of these precursors, in addition to their natural sources, tropospheric O_3 levels have doubled since pre-industrial times, as shown in Figure 6.7.

Because of the high chemical reactivity of O_3, it has not proved possible to recover pre-historic O_3 levels from ice cores as it has for CO_2, CH_4 and N_2O. Instrumental O_3 measurements began in the

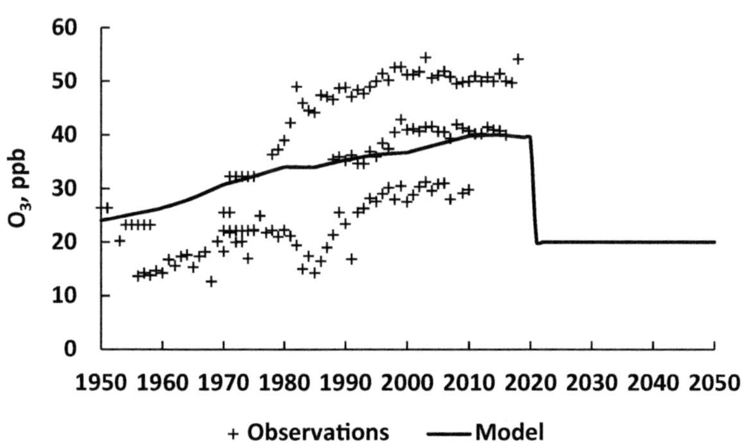

Figure 6.7 The time history of tropospheric O_3 from 1950–2020 as predicted by our global climate model.

1920s but these measurements were sporadic and unreliable. Reliable O_3 measurements began in the 1950s with relatively poor coverage and in the 1970s with reasonable coverage and reliable quality as shown in Figure 6.7. The simple global climate model predictions are also shown in the figure and demonstrate a reasonable fit against the available observations. The figure demonstrates how tropospheric O_3 levels have doubled since pre-industrial times and then fallen back to pre-industrial levels with an assumed complete cessation of man-made O_3 precursor emissions in 2020.

The global distribution of tropospheric O_3 provides a background level over the northern hemisphere continents on top of which regional O_3 episodes sit. The rise in global O_3 levels illustrated in Figure 6.7 has led to a rise in regional O_3 levels over North America and Europe that has partially offset air quality improvements due to regional pollution emission controls. It has now been realised in policy circles that there is a global dimension to regional O_3 and that future policies within North America and Europe will need to take intercontinental long-range O_3 transport fully into account.

RADIATIVE FORCING OF THE GLOBAL CLIMATE SYSTEM

We have, using our global climate model, quantified the extent of atmospheric composition change since pre-industrial times involving the main greenhouse gases: CO_2, CH_4, N_2O, halocarbons and O_3. To move down the global change framework diagram in Figure 6.1, we need to convert these changes in atmospheric composition into radiative forcing. To do this we use the radiative efficiencies of each greenhouse gas to relate atmospheric composition change in ppm since pre-industrial times to radiative forcing in W m^{-2}. The radiative efficiencies of the major greenhouse gases are calculated from the physical properties of each greenhouse gas, specifically their infrared spectra, and are tabulated in Table 6.1 below.

Radiative efficiencies span a range of ten thousand or more from CO_2 to CFC-12 and more than make up for the vast range in the atmospheric composition changes since pre-industrial times, from the

Table 6.1 Radiative efficiencies in W m^{-2} ppm^{-1} for each of the major greenhouse gases.

Greenhouse gas	*Radiative efficiency, W m^{-2} ppm^{-1}*
CO_2	0.0137
CH_4	0.363
N_2O	0.30
O_3	2.57
CFC-12	320
CFC-11	260
HCFC-22	210

CFCs in parts per trillion to CO_2 in parts per million. Putting together all the contributions to radiative forcing from the greenhouse gases, we arrive at the time history of the radiative forcing shown in Figure 6.8.

In arriving at this time history, we have assumed that the emissions of the greenhouse gases and aerosol precursors have been maintained at their 2020 levels. This is an illustrative scenario that allows the predictions of future global climate change to be made under the assumption that growth in emissions has been halted in 2020. This is not a realistic scenario but it is straightforward to understand and allows us to make some simple predictions. We use more realistic scenarios in Chapter 7.

The time history of radiative forcing from 1751 through to 2020 shown in Figure 6.8, illustrates how CO_2 contributes the greatest to radiative forcing, with CH_4, O_3, halocarbons and then N_2O next in importance. In the eighteenth and nineteenth centuries, CH_4 forcing dominated over CO_2 and all other man-made greenhouse gases. This is because the movement of the population into the towns and cities to work in the factories of the industrial revolution separated a large part of the population from sources of food. Agricultural practices were forced to change so that they could supply the increasing requirements of food for the urban and industrial centres. This intensification of agriculture led to dramatically increased CH_4 emissions and this intensification has been maintained through to the present day.

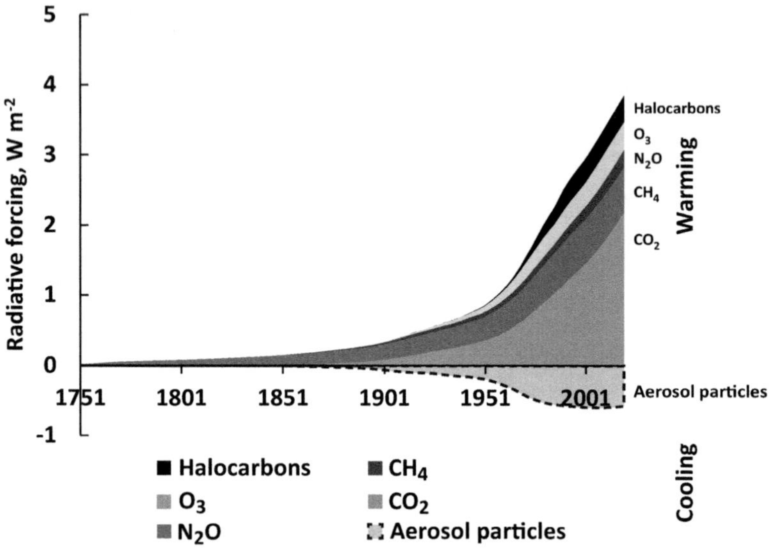

Figure 6.8 The time history of radiative forcing from CO_2, CH_4, N_2O, tropospheric ozone, halocarbons and aerosols.

AEROSOLS

Several different types of aerosol particles are present in the atmosphere and each has a different impact on radiative forcing. Most aerosol particles scatter and or absorb incoming solar short-wave radiation and out-going long-wave infrared radiation. A range of distinct components can be identified in driving the interaction between aerosol particles and radiation, including: ammonium sulphate, ammonium nitrate, black carbon from fossil fuel and biomass burning, organic particles from biomass burning, secondary organic and mineral dust particles. Aerosol particles also may increase the number of cloud condensation nuclei and this in turn may increase cloud albedo and the reflectivity of Earth's surface. In this way, aerosol particles induce a negative radiative forcing (cooling) and act to counteract partially the positive radiative forcing (warming) from the greenhouse gases, as shown in Figure 6.8.

PREDICTING THE RISE IN GLOBAL TEMPERATURE

The greenhouse gases and aerosol particles provide a forcing to the global climate system which is presented in Figure 6.8. It is a huge scientific challenge to predict the global temperature change that will result from this forcing. The response of the climate system will be delayed by the large heat capacity of the oceans and the ocean currents and their slow responses, will drive geographic variations in the temperature response. It is therefore crucial to have a coupled description of the atmosphere and the oceans if we are to predict the global climate change that will result from the radiative forcing induced by the atmospheric composition change. The coupled atmosphere-ocean models used to predict future man-made climate change are called global climate models (GCMs).

Initially, GCMs were extensions of the weather prediction models that were in general use throughout the world's meteorological services. Radiation schemes were added to handle clouds, aerosol particles and radiation. Ocean models were coupled to the atmosphere, initially in the form of slab or mixed layer models, but then increasingly complex ocean circulation models were added. Atmospheric chemistry and biogeochemistry were finally included to produce the breath-takingly complex Earth system models (ESMs). ESMs are now the standard tools for the assessment of present and future climate change and are making a huge contribution to climate change policy development and assessment.

For this monograph, we have assembled a global climate model using a mixed layer ocean model in which the ocean is represented by a wet surface with the heat capacity of the mixed or surface layer of the oceans. Details of this climate model are given in Appendix 4. In this way, we can represent the global temperature response to the imposed radiative forcing using an assumed heat capacity of the oceans.

Using our climate model, we can predict the rise in global temperature since pre-industrial times that has resulted from the man-made emissions of each of the greenhouse gases and aerosols. Figure 6.9 presents the predictions for CO_2 alone, the most important greenhouse gas, assuming that 2020 emissions are held constant

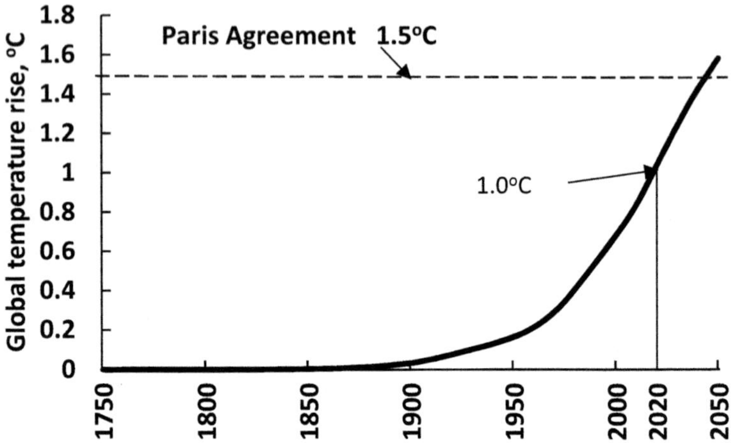

Figure 6.9 Global temperature rise since pre-industrial times predicted by the global climate model for CO_2 alone, showing the current temperature rise of 1.0°C and the Paris Agreement target of 1.5°C.

thereafter. The predicted temperature rise passed 0.5°C in the year 1986, 1.0°C in 2020 and is predicted to reach 1.6°C by 2050.

When the nations of the world signed the Paris Agreement on 12 December 2015, our climate model predicted that there was only limited capacity left available to the future climate system if the global temperature rise from CO_2 emissions alone is to be kept to below 1.5°C. By the time the nations had left the COP 26 meeting of the parties to the UNFCCC in Glasgow in November 2021, we have used up one-fifth of this available capacity. The entire capacity will be used up by the end of 2037, on present emissions but the world will be nowhere near zero-carbon.

Also shown in Figure 6.9 is the Paris Agreement target of 1.5°C rise in global temperature since pre-industrial times. The countries of the world agreed to take whatever measures they could to keep the rise in global temperature to below the 1.5°C target. Figure 6.9 shows that if man-made CO_2 emissions are held constant at their 2020 levels, then the Paris Agreement target will be reached in 2045 based on carbon dioxide alone. It therefore appears that there are twenty years or less in which to take the necessary measures to meet the challenge of the Paris Agreement target. However, allowance must be made for the other greenhouse gases and aerosols and this makes the challenge more difficult.

The global temperature rises predicted for CH_4 and tropospheric O_3, the second and third most important man-made greenhouse gases after CO_2, are shown in Figure 6.10. The temperature rise from CH_4 dominates that from all other greenhouse gases during the eighteenth, nineteenth and first half of the twentieth centuries. CH_4 emissions grew quickly during the Industrial Revolution because of the rapid growth in coal mining and from the intensification of agriculture following the movement of the population from the countryside into the industrial factories. The temperature rise from O_3 appears two centuries later than that from CH_4, following the growth in the use of petroleum-based fuels for transport.

Together CH_4 and O_3 produce a rise of about one half of that from CO_2 alone. However, they are both relatively short-lived greenhouse gases and so they can play a potentially vital role in meeting the challenge of the Paris Agreement target because they

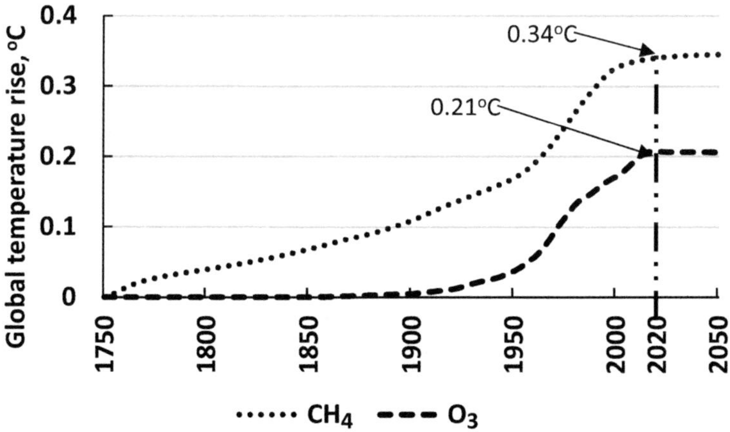

Figure 6.10 Global temperature rise since pre-industrial times predicted by our global climate model for CH_4 and tropospheric O_3, showing the current temperature rises of 0.34°C and 0.21°C, respectively.

will respond quickly to control measures and emission reductions aimed at maintaining the global temperature rise below 1.5°C.

N_2O appears as a significant greenhouse gas in the middle of the nineteenth century and contributes a global temperature rise of about 0.12°C, as shown in Figure 6.11. N_2O is an extremely long-lived greenhouse gas and so its contribution to global warming increases steadily to 2050 even if emissions are held constant at 2020 levels. It will be exceedingly difficult to bring down the contribution to the global temperature rise from N_2O. Indeed, it will be a considerable challenge to merely stabilise its contribution before 2050 and so help in meeting the Paris Agreement target, rather than making the target more difficult to meet.

In contrast, halocarbons have contributed to the rise in global temperature only since the middle of the twentieth century, see Figure 6.11. The halocarbons are a complex mixture of chlorinated, fluorinated and brominated organic compounds, each with its own atmospheric emission, atmospheric lifetime and radiative efficiency. Two large contributions have arisen from CFC-11 and CFC-12 over the period from 1950–2000 from their use as aerosol propellants, foam blowing agents,

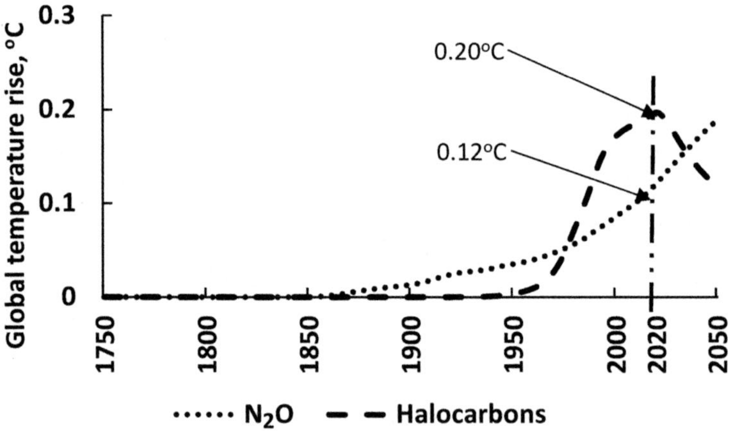

Figure 6.11 Global temperature rise since pre-industrial times predicted by the global climate model for N_2O and the halocarbons, showing the current temperature rises of 0.12°C and 0.20°C, respectively.

refrigerants and air-conditioning fluids. Since 2000, contributions have largely arisen from their HCFC and HFC replacements.

Together, N_2O and the halocarbons contribute 0.32°C to global warming (see Figure 6.11), which is about one-third of the contribution from CO_2. It will therefore be essential to consider carefully the potential role of emission control measures for N_2O and halocarbons in meeting the global climate change challenge.

Aerosols exert a cooling influence on the global climate system as shown in Figure 6.12. This cooling began in the 1850s and increased rapidly during the twentieth century, driven by the emissions of aerosol precursors from coal burning power stations. Our climate model predicts a decrease of –0.46°C in the global atmospheric temperature for 2020 due to aerosols, see Figure 6.12.

This global climate cooling presented in Figure 6.12 has been exceedingly important because it has partially counterbalanced the warming from the greenhouse gases, particularly over the period from the 1950s to the present date. Indeed, it is currently offsetting almost one-half of the global warming from CO_2 alone. This cooling has therefore disguised and hidden much of the

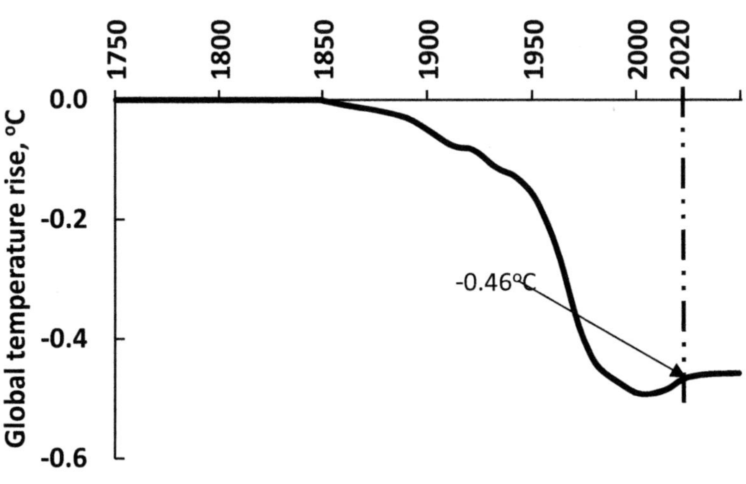

Figure 6.12 Global temperature decrease since pre-industrial times predicted for aerosols, showing the current decrease of –0.46°C.

greenhouse gas warming and lulled us into a false sense of security. This is a hugely important aspect of the global climate challenge and we need to understand urgently whether aerosols will continue to counterbalance the greenhouse gas warming.

We have summed the global warming contributions from the greenhouse gases in Figures 6.9 to 6.11 and subtracted off the cooling from the aerosols in Figure 6.12, using the predictions from the climate model. The resultant global temperature rise is compared to the observed rise in global temperature rise since 1880, reported relative to the 1880–1920 average, by James Hansen of the Goddard Institute for Space Studies in New York, as shown in Figure 6.13 below. The agreement between the climate model and the GISS temperature record is fairly close throughout the 1880–2020 period. Indeed, the climate model gives a global temperature rise for 2020 of +1.45°C, with a likely range of 1.15°C to 1.75°C, compared to +1.29°C in the GISS temperature record. Without the counterbalancing effect of the aerosol cooling, our climate model would greatly overestimate the observed global temperature rise throughout the 1880–2020 period. As it is, our climate model overestimates the observed 2020 rise by about +10%, but this is within the margin of error in our climate model as shown by the error bars. This comparison provides some reassurance that model performance is adequate for our present purpose.

The details of our global climate model are laid out in Appendix 4, with specific exercises addressing CO_2, CH_4, N_2O, O_3, halocarbons and aerosols in Exercises A4.1 to A4.6, respectively. In building our climate model, we have, out of necessity, made a large number of assumptions and simplifications. Consequently, there are large uncertainties in its predictions which we need to quantify. A Monte Carlo approach to quantifying these uncertainties has been employed. With this approach, we have identified nineteen model input parameters which must be set in the model and rerun the model 10,000 times with slightly different randomly sampled values. The error bars in Figure 6.13 show the likely range in the global temperature found in each year from 1880–2020.

A consideration of the error bars demonstrates the large degree of uncertainty present in the predictions of our climate model. The agreement found between the model and the observed surface

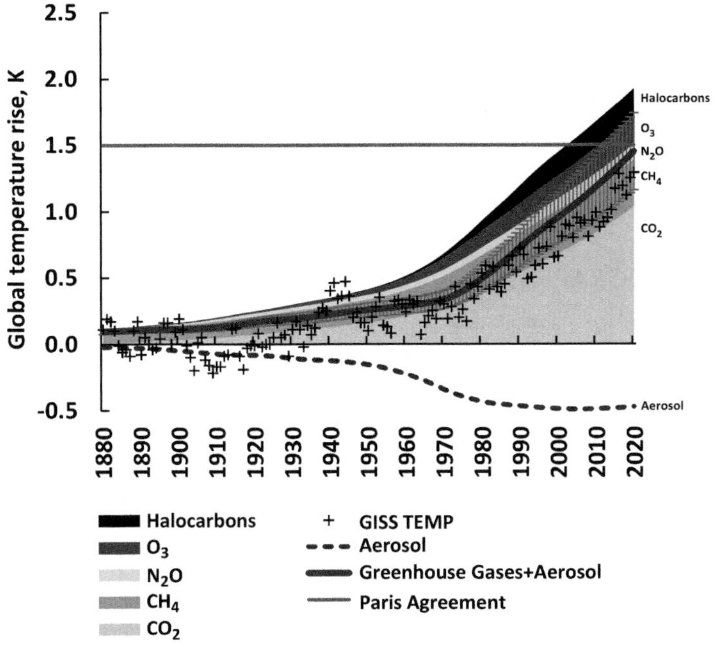

Figure 6.13 Global climate model predictions of the rise in global temperature between 1880 and 2020 driven by greenhouse gases and aerosols, compared with the observations in the Goddard Institute for Space Sciences (GISS) global temperature record, relative to the 1880–1920 average. The error bars show the likely range in the model predicted global temperature rise.

temperature record is, therefore, more a testament to the choice of the 'best estimate' input parameters and less to the adequacy of the scientific underpinning and its treatment of the complex processes involved. We have had to make substantial simplifications and approximations in building our climate model but at each stage we have carefully compared our model predictions with observations. This has reassured us that the model performance is adequate for our purpose.

Our global climate model predictions anticipate reaching the Paris Agreement 1.5°C target already in 2020s. This means that there is, in fact, little margin for delay and that the opportunity to keep the global temperature rise to 1.5°C thereby avoiding dangerous climate change, may well have passed us by already. It seems that some level of dangerous climate change is inevitable, should greenhouse gas emissions continue at present levels.

BIBLIOGRAPHY

BP, 2020. *BP statistical review of world energy 2020*, 69th edition. London: BP.

Duncombe, J., 2021. What five graphs from the UN climate report reveal about our path to halting climate change. https://eos.org/articles/what-five-graphs-from-the-u-n-climate-report-reveal-about-our-path-to-halting-climate-change.

Fiore, A. M., West, J.J., Horowitz, L.W., Naik, V., Schwarzkopf, M.D., 2008. Characterizing the tropospheric ozone response to methane emissions controls and the benefits to climate and air quality. *Journal of Geophysical Research Atmospheres* 113, D08307. (doi:10.1029/2007JD009162)

Hamilton, C., 2010. *Requiem for a species: Why we resist the truth about climate change*. Abingdon: Routledge.

Hansen, J., Sato, M., 2021. The world has cooled off – so what's the significance? www.columbia.edu/~mhs119/Temperature/Emails/April2021.pdf.

Harmsen, M.*et al.*, 2019. Taking some heat off the NDCs? The limited potential of additional short-lived climate forcers' mitigation. *Climate Change* 163, 1443–1461.

Hoglund-Isaksson, L., Gomez-Sanabria, A., Klimont, Z., Rafaj, P., Schopp, W., 2020. Technical potentials and cost for reducing global anthropogenic methane emissions in the 2050 timeframe – results from the GAINS model. *Environmental Research Communications* 2, 025004.

Intergovernmental Panel on Climate Change, 1990. *Climate change: The IPCC scientific assessment (1990)*. Cambridge: Cambridge University Press.

Intergovernmental Panel on Climate Change, 2005. *IPCC special report on carbon dioxide capture and storage*. Cambridge: Cambridge University Press.

Intergovernmental Panel on Climate Change, 2021. *Climate change 2021: The physical science basis. Summary for policymakers*. Cambridge: Cambridge University Press.

International Energy Agency, 2020. Methane tracker 2020. www.iea.org/reports/methane-tracker–2020.

International Energy Agency, 2021. Global energy review CO_2 emissions in 2020. www.iea.org/articles/global-energy-review-co2-emissions-in–2020.

Mason, J., 1992. *The greenhouse effect and global warming: The Linacre Lectures 1990–1991*. Oxford: Oxford University Press.

Monks, P.S., *et al*., 2015. Tropospheric ozone and its precursors from the urban to the global scale from air quality to short-lived climate forcer. *Atmospheric Chemistry and Physics* 15.

United Nations Environment Programme, 2017. *The emissions gap report 2017*. Nairobi, Kenya: UNEP.

United Nations Framework Convention on Climate Change, 2021. *Glasgow climate pact*, advanced unedited version. Bonn, Germany: UNFCCC.

Warszawski, L., *et al*., 2021. All options, not silver bullets, needed to limit global warming to 1.5°C: a scenario appraisal. *Environmental Research Letters* 16, 064037.

7

FUTURE SCENARIOS FOR AIR QUALITY AND CLIMATE CHANGE

OVERVIEW

In this chapter, we will use our global climate model to check out the effectiveness of some future scenarios in terms of reducing global warming. In making these assessments, it is usual to start by defining a base case scenario so that the impact of the measures can be seen more clearly. We have chosen a 'no coal' scenario as our base case. In this scenario, coal use is phased out by 2030 and with it the regional pollutant emissions of SO_2 and NO_x. This is a reasonable base case scenario because coal use in power stations is already being phased out in favour of natural gas and renewables.

Although there is a reduction in global warming greenhouse gas emissions, there is necessarily a reduction in aerosol precursor emissions and in global cooling aerosols in the 'no coal' scenario. There is also a reduction in global warming ozone and an increase in global warming methane because of the impact of NO_x reductions on atmospheric chemistry. The overall effect of the 'no coal' scenario is shown in Figure 7.1.

In the 'no coal' scenario, global temperatures, relative to the 1880–1920 average, are expected to climb past the Paris Agreement target of 1.5°C during the early 2020s and the 2°C target by 2033, see Figure 7.1. These predictions show that the temperature rise continues through the 2040s passing 2.2°C in 2050. Global temperature rises of this magnitude will be disastrous for the planet.

DOI: 10.4324/9781003293132-7

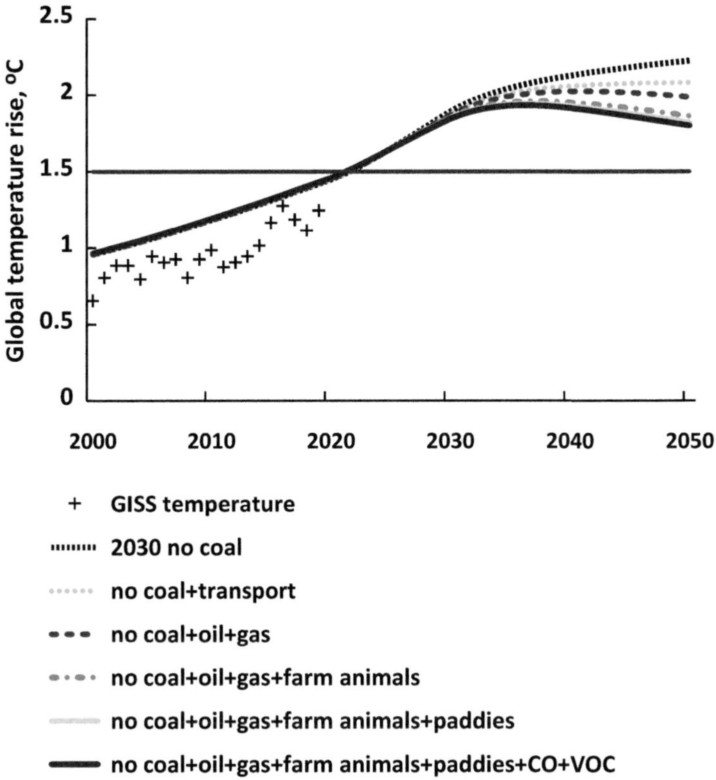

Figure 7.1 Global temperature predictions out to 2050 with for various scenarios involving future emissions of regional and global air pollutants.

The apparent acceleration in the growth of the global temperature during the 2020s is caused by the reduction in SO_2 emissions that act as aerosol precursors. Hence aerosol cooling declines quickly. This aerosol cooling has partially hidden the true scale of the global warming commitment that has built up, and has lulled policy-makers into a false sense of security.

Armed with this 'no coal' base case we will examine some of the measures that have been proposed and discussed for the control and reduction of the emissions of CO_2 and CH_4 on the global temperature predictions from our global climate model. These temperature rise predictions are also shown in Figure 7.1.

The measures are:

- Replacing coal-fired power stations by renewables and nuclear by 2030. This is the 'no coal' baseline and reduces CO_2 emissions by 42% of the 2020 total. This scenario also leads to reductions in the emissions of the regional scale air pollutants such as SO_2 and NO_x.
- Eliminating all gasoline and diesel vehicles by 2040, replaced by electric- or hydrogen-power. This is the 'no transport' scenario. It also leads to reductions in the emissions of the regional scale air pollutants such as CO, NO_x and VOCs.
- All CO_2 emissions from natural gas use have been eliminated by 2050 by assuming all industrial plants are clean and domestic heating is by hydrogen and solar power. This is the 'no coal + oil + gas' option. There are also reductions in the emissions of CH_4 from natural gas production, transmission and distribution. Additionally, methane emissions from waste disposal have been captured and used as a replacement for natural gas.
- All CH_4 emissions from farm animals related to human consumption, mainly intensive beef and dairy farming, have been eliminated by changes in diet, husbandry and manure management by 2040. This is the 'no coal + oil + gas + farm animals' scenario.
- All CH_4 emissions from rice paddies have been eliminated by changes in agricultural practices by 2040. This is the 'no coal + oil + gas + farm animals + paddies' scenario.
- All ozone CO and VOC precursor emissions from sources other than land transport have been eliminated by changing to hydrogen to make iron and steel (CO) and tight VOC emission controls in the chemical industry and substitution of low reactivity VOCs in solvents and household chemicals.

In this way, we have three scenarios addressing CO_2 emissions largely, two scenarios addressing CH_4 emissions largely and one scenario addressing ozone precursor emissions.

When these scenarios are fed into the global climate model, the predicted global temperature rises throughout the first half of the twenty-first century are found as shown in Figure 7.1. Because of the long atmospheric lifetime of CO_2 and the time lags in the global climate system, the impacts of the three CO_2 emission reduction scenarios are not felt until the 2030s, despite their implementation beginning in 2020. The five scenarios make little difference to the date when the Paris Agreement 1.5°C target is reached. This date shifts by about one year between the least and most ambitious of the scenarios. The scenarios, however, cause the global temperature rise above the 1880–1920 average to reach a maximum and begin to decline, behaviour not seen in the 'no-coal' baseline, see Figure 7.1. In the least ambitious scenario, the maximum in the global temperature rise is just below 2.0°C in the early 2050s. This maximum is reduced to about 1.9°C and occurs 14 years earlier in 2036 or thereabouts in the most ambitious scenario.

The five greenhouse gas emission reduction scenarios will exert a dramatic impact on the predicted rise in global temperature and will, ultimately, halt this increase. However, even with the most ambitious scenario, entailing rapid phase-out of all fossil fuel use, timescales are long and it will take upwards of 15–20 years for global warming to peak. By this time, the Paris Agreement target will have been passed many years previously and it will have taken further decades for the global temperatures to decline back and fall below it.

We are now in a position to be able to assess the impact of the measures taken to meet the challenges of global climate change on regional air quality. Accordingly, we have fed the global climate model predictions for future CH_4 and tropospheric O_3 into a regional O_3 air quality model, together with the historic and future European regional CH_4, CO, NO_x and VOC emissions. In Figure 7.2 we show the observed annual maximum 8-hour mean O_3 levels for 12 rural UK monitoring stations (open circles). Also shown are the regional model predictions for O_3 using these historic European regional emissions up to 2014 and held constant thereafter (solid black line). Episodic peak O_3 levels decrease in response to reductions in European regional emission reductions in agreement with the observations, but do not meet the WHO ozone air quality guideline value of 50 ppb (grey line). Implementation of the 'no coal- and gas-fired power stations' scenario envisaged an additional 25% reduction in European regional NO_x emissions on 2014 levels by 2040, which using our regional air quality model, points to a reduction in episodic

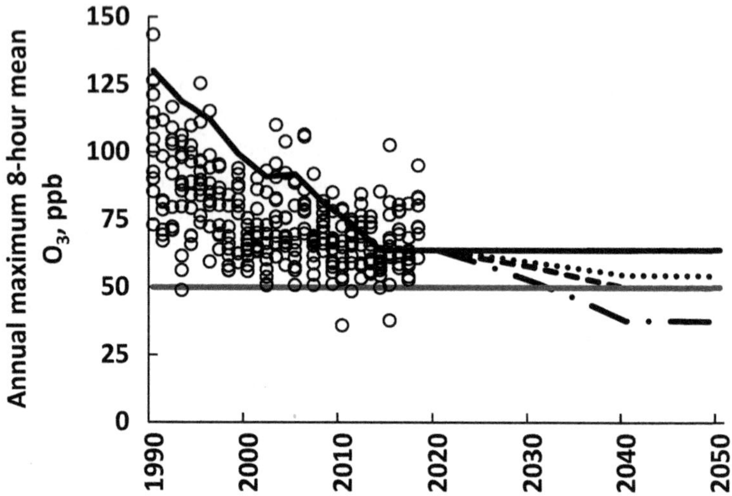

Figure 7.2 Time history of episodic peak O_3 level in rural areas of the UK from observations (open circles) and in a regional O_3 air quality model for 1990–2014 and model predictions out to 2050 for a 'no coal- and gas-fired power station' scenario (dotted line), 'no coal + transport' scenario (dashed line) and 'no coal + oil + gas' scenario (dashed and dotted line). Also shown is the WHO 2021 air quality guideline value for O_3 (grey line).

peak ozone from just under 64 ppb down to 54 ppb (dotted line) which is not quite enough to meet the WHO 2021 ozone air quality guideline. In the 'no coal + transport' scenario, measures were outlined to phase-out coal-fired power stations and road transport fuel but leave natural gas-fired power stations alone. These measures should lead to a halving in European NO_x emissions and 12% reduction in European VOC emissions by 2040, which should allow the WHO air quality guideline to be met during the 2040s (dashed line) with little margin of error.

The 'no coal + oil + gas' scenario envisaged phasing out coal, natural gas and road transport by 2040. Full implementation of this scenario across Europe should reduce regional NO_x emissions by just over 60% and VOC emissions by about 12%. Episodic peak ozone levels should fall to below 40 ppb, well below the WHO air quality guideline, (dotted and dashed line) representing a full 'win' for regional air quality. Although emissions of NO_x, CO and VOCs are substantially reduced in this scenario, significant regional emissions remain in 2040 arising largely from industrial NO_x and VOC sources, together with significant solvent emissions of VOCs. Because the WHO ozone air quality guideline has been achieved in this scenario, there appears to be little merit in tackling the residual VOC emissions from the industrial and solvents sectors and the residual industrial NO_x emissions.

We have shown that these scenarios improve air quality dramatically. Ozone, nitrogen oxides, VOCs and carbon monoxide will become pollutants of the past in less than 20 years.

Phasing out agricultural CH_4 emissions, in Europe leads to an overall 46% reduction in man-made methane emissions. When these methane reductions are implemented in the regional model, on top of the 'no coal + transport' scenario reductions, episodic peak ozone levels drop by a further 0.8 ppb. The impact is slight as regional O_3 predictions post-2014 are dominated by NO_x levels and the greatest reductions are found in scenarios with the deepest cuts in NO_x emissions on the local, regional and global scales. At all scales, photochemical ozone production appears to be highly NO_x-limited.

Reducing O_3 and CH_4 today gives an almost immediate effect on global warming reduction as we have seen in Figure 7.2. It remains for governments worldwide to cooperate internationally in the measurement and control of O_3 and CH_4, and reap the benefits of improving regional air quality and reducing global warming in a short time.

PHASE-OUT OF COAL- AND NATURAL GAS-FIRED POWER STATIONS AND ROAD TRANSPORT WITH REDUCTIONS IN AGRICULTURAL METHANE EMISSIONS AND GLOBAL-SCALE REFORESTATION

Nature-based solutions are built upon a number of interventions and measures which seek to draw down CO_2 from the atmosphere and store it in the biosphere. Some examples are:

- Existing natural CO_2 sinks in forests, mangroves, grasslands and sea-grasses should be kept intact.

- A moratorium on deforestation.
- Restoration of degraded forests, grasslands and wetlands.
- Large-scale reforestation in sub-tropics and tropics.
- Sustainable use of existing sub-tropical, temperate and boreal forests.
- Forest ecosystem restoration.
- Land-based sequestration in temperate, boreal, sub-tropical and tropical forests.

To illustrate the potential power of land-based solutions, we have built a reforestation scenario in which a global tree-planting strategy is adopted which leads to the planting of a million trees globally in each year from 2020, on top of the most ambitious CO_2 and CH_4 scenario from Figure 7.1. It is difficult to estimate precisely how much CO_2 would be sequestered in such a simple scenario because the amount would vary depending on the age of the trees as they grow, where they grow and species of the trees themselves. To provide an indication of what could be achieved by a simple global tree-planting strategy, we use mid-range estimates from the Woodland Carbon Code of the UK Forestry Commission.

Sequestration models point to CO_2 uptake being low in the first few years as new forest stores little or no CO_2 as the trees are only beginning to grow. By the time the trees are 20 years old or so, the forest canopy closes and CO_2 uptake is at its maximum. As the forest ages, the rate of uptake slows as new tree branches form but old branches decay until the rate of uptake levels out after about 70 years.

When the CO_2 sequestration rates are calculated for each year after 2020 in our 1 million trees per year strategy and are fed into our climate model, we can estimate the impact of the nature-based CO_2 uptake on the global temperature rise, see Figure 7.3.

The CO_2 sequestration by the reforestation scenario halts the rise in global temperature, causes a levelling off at 1.9°C several years earlier than the most ambitious greenhouse gas emission scenario in Figure 7.1. Ultimately, the reforestation scenario brings about a decline in global warming, bringing it below the Paris Agreement 1.5°C target in 2055.

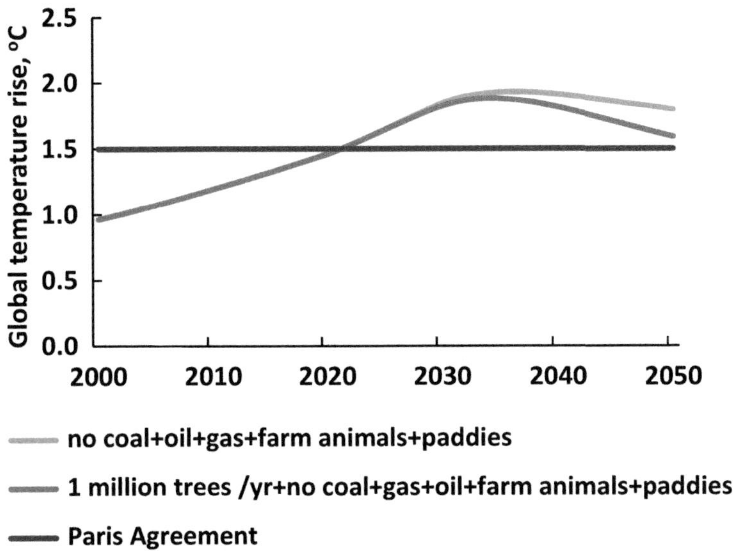

Figure 7.3 Climate model predictions for the global reforestation scenario on top of the most ambitious CO_2 and CH_4 emission reduction scenarios.

If these greenhouse gas emission and reforestation scenarios are adopted by countries worldwide, we will have protected our planet from dangerous climate change and eliminated serious air pollution.

8

THE WAY FORWARD

These fundamental things must be simple.

– Lord Rutherford

OVERVIEW

We summarise in this chapter, the important steps necessary to improve air quality and reduce global warming.

It is essential to have legislation on indoor air quality, particularly on $PM_{2.5}$ and carcinogenic VOCs which should be set at WHO outdoor air quality standards.

Urban and regional air quality will greatly benefit from the replacement of gasoline and diesel passenger cars by electric vehicles and heavy-duty vehicles should give way to hydrogen technology.

While the elimination of coal- and gas-fired power stations will improve air quality, atmospheric CO_2 levels will take a long time to respond and it will take many decades for global climate change to show a real benefit. Reduction in methane from all sources will be a win–win policy, especially when combined with ozone reduction. However, it is when we have a scenario of reforestation combined with reductions in emissions of all greenhouse gases that we can realise an improvement in climate change which will enable the planet to limit global warming to 1.9°C.

Over the next 30 years, air quality will dramatically improve and global warming will remain a challenge. Yet, we will succeed in reversing global warming and bring it down to 1.5°C by 2055.

DOI: 10.4324/9781003293132-8

The great physicist, Lord Rutherford, always ended his lectures with this statement, 'these fundamental things must be simple'. This is our approach in improving air quality and reducing global warming, and we will achieve these aims step by step.

Teamwork and determination are keys to success. Our team consists of scientists, research institutions, local and central governments, industry and legislators working together on achieving a central objective in a given time frame.

Let us begin with the improvement of indoor air quality as discussed in Chapter 3. We have seen that there is a real need for indoor air quality improvement in the reduction of $PM_{2.5}$ and certain VOC species. Yet, there is little legislation and we look to legislators in each country to adopt the WHO 2021 air quality guidelines and criteria for $PM_{2.5}$ and carcinogenic VOCs as for outdoor air quality as soon as possible. Continuous monitoring of these pollutants in buildings and offices will lead to a steady improvement. It then becomes important for businesses to improve both ventilation and air conditioning systems. Time starts now!

At the same time, we must encourage the chemical and solvent industries to improve both the air quality of VOCs by further reduction and substitution of certain species. These same industries are working to provide a labelling system of their products in terms of health and safety.

URBAN AIR QUALITY

The best way to tackle urban air quality is for each local authority to establish an emissions inventory for the region under study. Specifically, we will be monitoring NO_x, $PM_{2.5}$ and CO_2 as a first step. During the Coronavirus lockdown in Europe in 2020, we saw that these pollutants were reduced by at least 30% in a few months in urban regions and blue skies became the norm. This was realised mainly by the dramatic reduction of diesel and gasoline road transport.

These reductions in NO_x, $PM_{2.5}$ and CO_2 will now be achieved without lockdown. We need to know traffic volumes and their distribution in terms of emissions and age-both passenger cars and heavy-duty vehicles and buses, as well as details of all industries and factories in the neighbourhood. One simple way of building a vehicle inventory is to characterise vehicles by engine size and age and so ascertain their emissions levels relevant to the year of construction. Industrialists can provide details of pollutants from each refinery, power station, incinerator and local factory, if only from a knowledge of published emissions factors. Each industry can be encouraged to monitor these emissions regularly, and so establish ownership of improving air quality. We can then build an inventory of all pollutants and their sources for each urban and surrounding region.

It then remains to apply emission controls to each sector ranging from new efficient plants to traffic control and modernisation.

Passenger cars and heavy-duty vehicles powered by gasoline and diesel fuel contribute nearly 25% of all CO_2 emissions in the atmosphere and 50% of all NO_x emissions. These vehicles need to be replaced by alternative technology over the next two decades in a careful cost-effective manner.

PASSENGER CARS

Motor manufacturers are well advanced in their design of electric passenger cars and construction of new gasoline and diesel engines are being phased out before 2030. The new generation of electric cars will have a longer travel range before refuelling and, given central government fiscal support, will be readily affordable.

As an example, we observed in Chapter 5 on regional air quality that some countries such as Poland have over 36% of passenger cars over 20 years old. Even if these vehicles are regularly maintained, they belong to a bygone era of exhaust emission norms. A first step is to provide a generous scrappage allowance paid for by central government, the European Commission and motor manufacturers. This not only reduces vehicle emissions substantially but improves the economy by providing new work in the automotive industry. Thereafter, the car distribution in each country can be modernised in stages by reducing the number of older vehicles and high polluting vehicles. $PM_{2.5}$ emissions from passenger cars will also be reduced as particles from diesel engine exhausts move into history.

We can also observe that many families have two or more cars, but after the Coronavirus pandemic, we move into a new way of life which no longer requires a multiplicity of passenger cars. There are currently over 239 million cars in China, 47 million in

Germany, 39 million in Italy, and 32 million in both France and the UK. Do we really need so many passenger car vehicles? Can we make better use of our efficient public transport?

It is realistic to predict that CO_2 emissions from the passenger car fleet will be halved in 2030, and could be reduced to zero by 2040. Given that gasoline and diesel car engines account for 30% of all NO_x emissions, their replacements by electric passenger cars will substantially reduce NO_x by 2030 and effectively eliminate NO_x from passenger cars by 2040. Indeed, these emissions improvements could well be achieved earlier given active encouragement by both governments and motor manufacturers for replacement of older diesel and gasoline powered passenger cars. As we have predicted in Chapter 5 on regional air quality, these reductions in NO_x will lead to substantial reductions in O_3 and enable the WHO air quality guideline value of 50 ppb for peak O_3 levels to be achieved.

HEAVY-DUTY VEHICLES AND BUSES

Motor manufacturers are already exploring hydrogen-powered heavy-duty vehicles as cleaner and lighter, and this has to be a way forward. One can envisage long distance travel before refuelling, and then service station refuelling will be as rapid as filling up with diesel or gasoline fuel. Energy companies are already paving the way for ease of hydrogen supply. Indeed, they are exploring the manufacture of hydrogen from natural gas.

There are different ways of manufacturing hydrogen, one being steam CH_4 reformation, which is energy intensive with energy companies capturing the CO_2 and effectively burying it below ground. Another method is electrolysis of water which is currently expensive, yet as wind generated electricity declines in cost, this method will prove more viable.

Scientists and engineers will ensure hydrogen automotive power is within our grasp, resulting in NO_x, $PM_{2.5}$ and CO_2 emissions being eliminated from heavy duty road transport by 2040 at the latest.

FUTURE AIRCRAFT

We recognise that commercial aviation emits 2.5% of all man-made carbon dioxide emissions and that large battery-powered aircraft are not a likely option in the future. We need an alternative to conventional kerosene if aircraft are to become carbon-neutral. Fortunately, there is a way forward which is being explored commercially. It is also possible that hydrogen powered aircraft may become available in the long term.

If water and electricity produced from wind farms are converted into hydrogen and the hydrogen combined with carbon dioxide to produce crude oil, then this in turn can be refined to make jet fuel. This synthetic kerosene is carbon-neutral because the resulting carbon dioxide released into the atmosphere is exactly the same as the amount taken out of the atmosphere to produce the fuel in the first place.

A modern plant has just opened in Germany to produce this synthetic kerosene and could soon become a major source which will become economically viable.

SHIPPING

Shipping continues to use heavy fuel oil, with the result that emissions of NO_x, SO_2 and particles remain substantial. More efficient combustion and pollution control devices will become the norm and it is anticipated that hydrogen fuel will be adopted in the longer term. All these developments will eventually lead to the establishment of 'green shipping corridors'.

POWER STATIONS, REFINERIES AND HEAVY INDUSTRIAL PLANTS

Our emissions inventory for the region also includes NO_x and CO_2 calculations from each industrial source. Coal-fired power stations are being replaced initially by natural gas which reduce CO_2 generated by 50%, and then by renewable energy sources which reduce CO_2 emissions to zero. Energy companies predict demand for oil to be reduced by 40% in 2030, which will impact on refinery operations. It is likely that many refineries will close and the capacity of others adjusted. These plants can be put to other uses, such as the manufacture of hydrogen. Heavy industrial plants have suffered during the coronavirus pandemic, and many will not return to full operations. Various scenarios can be factored into our emissions inventory following discussions with industry as we move into the future. The overall result will be substantial reductions of NO_x and CO_2 from heavy industry by 2030 and beyond.

Urban air quality is about to improve dramatically, and we can look forward to clear blue skies and quieter and cleaner cities. In reality, zero carbon emissions can be achieved before 2050, provided all countries and all teams working on emissions reduction co-operate closely. We are moving onwards and upwards.

REGIONAL AIR QUALITY

In Chapter 5, we looked in detail at regional air quality. Individual countries are responding to the challenges of reducing the main pollutants in a spirit of co-operation as seen with the United Nations Economic Commission for Europe protocols for NO_x and VOCs. These reductions of NO_x and VOCs will lead to O_3 levels falling and the WHO air quality guideline for peak O_3 levels of 50 ppb being achieved by the mid-2030s.

China emulates EU emissions standards as it has human health and safety as a key priority. This can be demonstrated by the manufacture of new cars in China which adhere to the latest EU norms. The energy policy in China is to ensure adequate energy supply to sustain economic growth. China consumes 70% of all coal used in Asia, almost 2 billion tonnes in 2019. There is no real evidence of changing to natural gas fired power stations, yet, other forms of power generation are in abundance with seven new large-scale nuclear reactors operational in 2019. There are over 150 power stations in China of which 54 are coal-fired, 14 are nuclear, 68 are hydroelectric, 10 are solar power generated, nine are powered by wind farms and one is tidal generated. Indeed, China is the largest producer in the world of hydroelectric, solar and wind power. As such, China has the potential and technology to change from coal fired power stations quickly given the political will.

Asia still relies heavily on coal combustion for electricity generation and would do well to follow the lead of China in moving to modern forms of energy generation.

The United States of America has switched to natural gas fired and wind power generated electricity, but still relies heavily on coal which is a political policy to protect employment. However, a new administration in Washington in 2021 is putting the improvement of the environment at the top of the agenda, and air pollution will decline in the next decade. The US Environmental Protection Agency is proactive.

Of course, we have seen that emissions from Asia are transported to Europe and beyond to create ozone and other regional pollutants. This is why we need to have increased co-operation between countries and governments. Yet, if all countries tackle their urban air quality, particularly in large cities and their surrounding regions, all pollutants including NO_x and VOCs will decline rapidly and less O_3 will be generated and transported.

GLOBAL WARMING AND CLIMATE CHANGE

FUTURE GREENHOUSE GAS EMISSION REDUCTIONS

In Chapter 4, we detailed how urban air quality would benefit from the major changes made in future fuel technology in road transport. Reductions in CO_2 emissions alone could be of the order of 25% by 2040, given a concerted effort by all determined to save the future of our planet.

Coal was the first source of energy which created the Industrial Revolution in 1750, and has served its purpose well. However, it remains the greatest source of CO_2 emissions relative to other fuels, notably natural gas and recently renewable sources. Coal combustion remains a major generation of CO_2, particularly in coal fired power stations. There is a need to switch to natural gas fired power stations and electrical energy generated by renewable sources such as wind, solar and nuclear power. The average emissions factor for CO_2 generated in a power station is 1.0 tonne MWh^{-1} from coal compared to 0.5 tonne MWh^{-1} from natural gas, thus halving the CO_2 by switching from coal to gas.

Coal fired power stations are changing to natural gas in the United States with energy related CO_2 emissions reduced by 140 million tonnes in 2019 corresponding to a 15% reduction in coal use in power generation. Natural gas is used to generate 37% of all electricity generation. The United States consumes over 90% of all coal used in North America – over 320,000 thousand tonnes in 2019. Much work needs to be done to reduce reliance on coal in the USA. The United States plans to achieve net zero carbon dioxide emissions by 2050 at the latest, which includes a pledge to double offshore wind power by 2030.

In Europe, there was a 5% drop in energy related CO_2 emissions in 2019, equating to 160 million tonnes. Electricity generated by coal-fired power plants has declined by 25%, with natural gas fired plants increased by 15%. Increasing use of renewable sources of electrical power continue to take over from coal fired generation. In Germany alone, there has been an 8% fall in CO_2 emissions, due to electricity generated from wind renewables increasing by 11%. Renewable energy generation now accounts for 40% of electricity generation in Germany.

In 2020, the United Kingdom generated more electricity from renewable sources including wind turbines, hydroelectric plants, solar panels and biomass plants than from fossil fuels. The total amounted to 42% with gas-fired electricity generation at 41% and nuclear plants generating 17%. The change from 2019 was due to the new wind farms which resulted in a 14% increase in wind power output year-on-year. Almost a quarter of UK power generation was from wind turbines in 2020.

Decarbonising power generation is very much the long-term aim of European countries by 2050. Will this be too late for our planet?

CO_2 emissions from energy generation in Japan fell by 4% in 2019, with reactors increasing in use for nuclear power. In India, electricity generation by coal combustion is falling, but otherwise, in Asia, reliance on coal is increasing, resulting in more than 50% of electricity generation. Clearly, the challenge is for Asia to reduce reliance on coal and move to natural gas and other renewable sources as soon as possible.

METHANE EMISSION REDUCTION

CH_4 emissions worldwide are currently about 600 million tonnes annually, an enormous amount, with China the principal anthropogenic source followed by India and the United States of America. CH_4 is emitted by agriculture from grazing animals especially cows, accounting for 30% of all emissions, and recent research is showing that these emissions can be reduced by different feedstocks, improved forage content and mitigation compounds. CH_4 reduction from farm animals will continue to progress. Rice paddies account for 15% of the total emissions, fossil fuel extraction and delivery account for 33% and waste disposal a further 18%.

We have seen that CH_4 accounts for an appreciable fraction of global warming and its reduction is critical. CH_4 can be removed from the atmosphere using zeolite which converts it into carbon dioxide. If we replaced all methane in this way to add to the carbon dioxide, we would reduce global warming by one-sixth.

We have also seen that CH_4 as natural gas can be used to manufacture hydrogen which is a fuel of the future. This is a new challenge for large energy companies as they move away from their dependence on oil.

Here, we have a gas which occurs naturally and which reacts to form ozone (O_3). After CO_2, it is the second most important greenhouse gas and its reduction constitutes a win–win scenario as both air quality and climate will distinctly benefit. CH_4 reduction is cost effective and does not depend on location, so climate and air quality improvements are achieved globally. O_3 reductions are almost directly proportional to CH_4 reductions, so CH_4 reductions make a significant impact on reducing background O_3 levels. This makes a distinct benefit to regional and global air quality. The reduction of CH_4 is making steady progress as new forms of power generation and domestic heating develop, so continuously reducing global warming from CH_4 sources. CH_4 emission reductions are a vital win–win scenario.

OZONE

Ozone (O_3) is a challenging pollutant which knows no international boundaries and is formed by the sunlight-driven photochemical reactions of VOCs and NO_x. It has a short lifetime of 20 days in the atmosphere and is an important greenhouse gas. Reducing O_3 is a win–win opportunity to improve human health and to reduce global warming. All nations have a major role to play particularly by reducing NO_x and methane emissions from all sources.

Transport is being modernised rapidly by phasing out internal combustion engine vehicles, particularly diesel transport with high NO_x emissions. We will see large increases of electric vehicles for passenger car transport in the next decade, and promotion of hydrogen fuel usage in heavy duty vehicles and buses. The recent

Coronavirus pandemic has led to a different lifestyle in developed countries and underlined the necessity to rely less on vehicle transport. Indeed, the number of car journeys and private car ownership has been reduced. Heavy industry which generates NO_x, is also in decline, notably oil refineries, as companies respond to reduced demand.

VOCs are being controlled by suppliers with more reactive hydrocarbons being replaced in the chemical industry. CH_4, a major source of O_3, is being reduced as we have indicated. We saw in Chapter 5 that O_3 peaks will fall steadily in line with NO_x reduction to bring levels down below the 50 ppb WHO ozone air quality guideline value by the mid-2030s.

THE DUAL CHALLENGES OF AIR QUALITY AND GLOBAL CLIMATE

We have identified three gases as having the greatest effect on man-made global warming namely CO_2, CH_4 and O_3. As discussed in Chapter 6, the three important factors are magnitude of the emissions, residence time and radiative efficiency. The great difference between them is the time taken to influence climate change, the atmospheric lifetime, with CO_2 taking over 100 years, CH_4 just over a decade, and O_3 some twenty days.

In Chapter 7, we have identified a number of measures that could be employed to address the dual challenges posed by air quality and global climate change. Reduction of CO_2 today will make a difference to future generations and will be achieved by minimising combustion processes especially in the automotive sector and electricity generation. We have seen that the largest source of CO_2 is from coal-fired power stations followed by oil-fired power stations. Asia remains the major user of coal and every encouragement by leading nations to reduce dependence on coal is essential. We also noted that should all emissions of CO_2 cease in 2020, this would lead to a very slow decline in CO_2 atmospheric levels, which is why pessimists say we may be to too late to stop climate change! In the meantime, we must reverse any government policy on deforestation. Trees absorb CO_2 and must be protected. We can and will reverse global warming and turn our attention to the other greenhouse gases to meet this challenge.

Aerosols such as sulphates, nitrates and secondary organic aerosols have been studied by many workers in the United States and

Europe as they contribute to the $PM_{2.5}$ pollution. It is now accepted that they reverse global warming as we can see clearly in Figure 6.8 in Chapter 6. Research workers need to highlight this benefit and legislators need to tackle the other sources of $PM_{2.5}$ in reducing particle pollution.

CH_4 reduction is a cost-effective way of reducing global warming both by its increase use as natural gas in heating, and power generation instead of coal. In agriculture, we have seen farm animals being fed on different feedstocks which lead to lower methane emissions. There is a real opportunity to use methane to generate hydrogen which is clearly an automotive fuel of the future, and so reduce methane further. CH_4 has a high radiative efficiency, and we have noted in Chapter 6 that reducing man-made CH_4 emissions to zero in 2020 would more than halve CH_4 concentrations in 10–20 years. Again, the policy of reducing methane needs to be clearly adopted by all nations and governments as a way forward in tackling climate change.

O_3 reduction has been the concern in air pollution since its identification by Haagen-Smit in 1952. Its transboundary nature led to blame matrices being generated by scientists in the 1970s, yet the UN Protocols identify the need for all countries and governments to work towards its reduction in a spirit of teamwork. O_3 is formed in the atmosphere by the photochemical reaction of NO_x and VOCs in sunlight and is a pollutant that knows no international boundary. Ownership of this pollutant is shared by all countries worldwide; the effect of reducing O_3 on climate change is almost immediate with O_3 having a high radiative efficiency, nearly 7 times more than CH_4. Reducing CH_4 emissions worldwide is essential in the drive to reduce O_3.

O_3 and CH_4 is the real win–win scenario for improving regional air quality and reducing global warming. This is the future challenge for all of us working to improve our environment.

NATURE-BASED SOLUTIONS

THE DANGER OF DEFORESTATION

There is much publicity at the moment covering the deforestation of the Amazon. CO_2 acts as a fertiliser for trees, with over 30% of all man-made CO_2 being absorbed by trees and forests, constituting a land sink. The subject of tree species and their uptake rates of CO_2 are still being researched. However, it is critical to plant trees and create forests whenever and wherever we can, to support the uptake of CO_2 by natural processes. Here, governments worldwide must be proactive in avoiding deforestation especially if they are determined to reduce global warming.

REFORESTATION

We have seen in Chapter 7 that a scenario of planting a million trees globally each year from 2020, in addition to all scenarios to reduce man-made emissions and reductions of CO_2 and CH_4, will make a real difference to global warming in the future, while further improving air quality. This is clearly shown in Figure 7.2. Indeed, although global warming will continue to rise until 2034, when it is predicted to peak at 1.9°C, it will fall below 1.5°C in 2055, in line with the Paris Agreement target of 2014. Figure 8.1

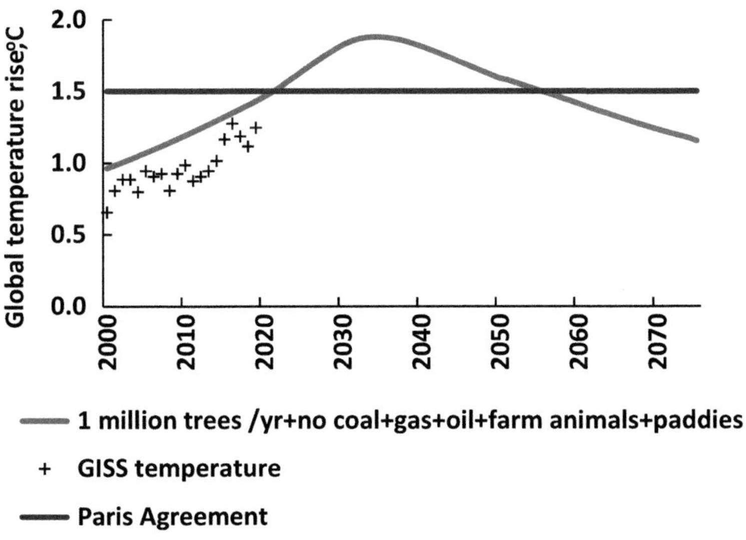

Figure 8.1 Predictions of global temperature rise extended to 2075 with the reforestation scenario and reduction of all man-made emissions.

illustrates how reforestation combined with reduction of man-made emissions will continue to reduce global warming as we move forward to 2075. By then global warming will be a concern of the past. Indeed, by 2075, the global increase in temperature will have fallen below 1.2°C.

COP-26 DECISIONS AND POTENTIAL REDUCTIONS IN GLOBAL WARMING

At the time of completing our monograph, world leaders have met in Glasgow in November 2021 to agree on a way forward to limit further increases in global warming. Here, we assess the potential reductions in global temperature if countries adopt the proposed COP-26 decisions.

- Coal is expected to be phased out as soon as possible. China and the United States of America need to lead the way but this will take time. We will assume optimistically that there is a phase-down in coal combustion beginning in 2030, and complete by 2050.
- Most motor manufacturers agree to produce emission-free vehicles by 2040. We will assume that half of all vehicles are emission-free in 2040 and the remainder by 2050.
- Anthropogenic methane emissions are expected to be reduced by 30% in the years ahead by all countries, including China and the United States of America. We will start phasing in the 30% reduction in methane emissions in 2030 and assume it is complete in 2050.
- Reforestation and the end of deforestation is expected to take place before 2030. We will start our reforestation scenario in 2020.

We now apply our model to calculate the global temperature increase, adopting the COP-26 measures, albeit assuming all countries play their full part. Figure 8.2 shows the results of implementing the COP-26 emission reductions. The COP-26 measures stop the global temperature rise from growing beyond 2050 and have the global temperature rise peaking at 2.3°C in the 2050s. It is only when the timescales for the emission reductions are advanced to 2021 and an ambitious reforestation program is

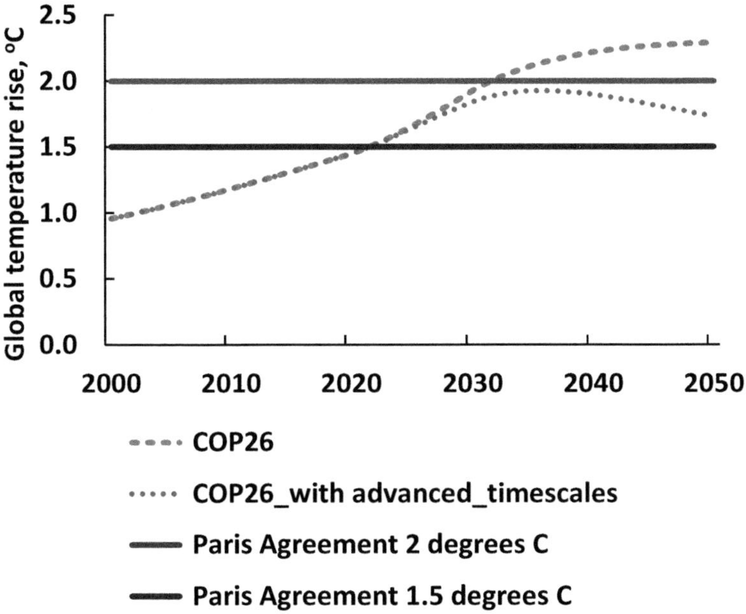

Figure 8.2 Global temperature rises in the COP-26 scenarios, together with the Paris Agreement targets.

implemented as part of the COP-26 measures, that the global temperature keeps below the Paris Agreement 2°C target.

COP-26 measures are a firm step in the right direction, yet, in the words of the Norfolk farmer, referring to mankind, 'If you would do what you could do, you'd do a deal better than you do do!'

The COP-26 emission reductions must be implemented immediately, without further delay. The UN Framework Convention on Climate Change will ultimately need to implement a phase-out of natural gas-fired power stations, take action on the use of petroleum-based fuels in international shipping and aviation and tackle methane emissions from agriculture. Finally, the UNFCCC will need to implement nature-based solutions by way of reforestation across the globe.

All the measures described in this chapter are necessary and sufficient to limit global temperature increases in line with the Paris

Agreement targets. Global warming will then have been controlled successfully by the mid-2030s, yet it will have taken longer than people anticipate. Reforestation, in particular, has the potential to reduce global warming cost-effectively and all countries can take action now. The 1.5°C target can easily be met by 2055. In the meantime, air quality has continued to improve and all pollutants including ground-level O_3, will fall within the 2021 WHO air quality guideline before 2050, and return to their pre-industrial levels well before 2100.

'These fundamental things must be simple.'

APPENDIX 1: INDOOR AIR QUALITY MODELS

In this appendix, we assemble and run indoor air quality models and apply them to some illustrative example situations. Most indoor air quality models are based on box models and this is where we begin.

We imagine that the indoor environment is a box, or more specifically a cuboid of floor area A (m^2), ceiling height h (m) and interior volume $V = Ah$ m^3. Outside air infiltrates the box because it is draughty, driven by the wind flow over the building, or forced by means of fans and ducts. When the flow of air into the indoor environment is balanced by the flow of air out, then a steady state or equilibrium exists. The inflow of air can be quantified in terms of the fraction of the box volume that is exchanged in unit time, that is, in terms of air changes per unit time. For a typical home, the inflow of air can be of the order of one air change per hour. In a draughty home then the inflow rate may be several times larger and in a well-sealed, draught-free house, several times less. This air change rate is sometimes called the ventilation rate. Ventilation may be natural, through the action of the wind, or forced by means of fans which suck air from the outside into the building. Outside air may often be polluted and so ventilation may bring pollutants into the indoor environment.

If the indoor environment contains a source of indoor air pollution, then the indoor concentration of the pollutant will build up whilst the source is operating and decay when it is switched off. This decay results from ventilation with cleaner air entering compared with the more polluted air leaving. Decay may also result from removal of the pollutant by chemical reactions within the indoor environment through uptake on furnishings and fabrics.

Let us assume that the indoor source emits S grams of pollutant per unit time and that the initial pollutant concentrations were zero when the source was switched on at time, $t = 0$. In each unit of time, Δt, the source will add $S.\Delta t$ grams into the box and this will lead to the pollutant concentration, c, rising by $S.\Delta t/V$, g m^{-3}, assuming that the pollutant is instantaneously mixed throughout the indoor environment. As air leaks out of the indoor environment, it takes with it some of the pollutant. If the ventilation or air exchange rate is a air changes per unit time, then the rate of loss of pollutant by ventilation is $aVc.\Delta t$. Assuming ventilation is in steady state then the outflow of air must be balanced by inflow. This inflow air may contain the background levels of the pollutant, c_b, and so adds $aVc_b.\Delta t$ to the contents of the box. If the pollutant is suspended particulate matter, then it may not be able to follow the labyrinth of draughty passages when entering the indoor environment. Penetration efficiencies, P, may not be unity for suspended particulate matter as they are for most gases and may lie in the range from 0.1 to 1. In this case, the pollutant inflow would be $aVPc_b.\Delta t$. Additionally, the pollutant may decay by uptake on soft furnishings and other surfaces with a decay rate of k per unit of time. The loss of pollutant mass per unit of time will be $Vkc.\Delta t$.

Applying the Law of Conservation of Matter at each point in time, the increase in the mass of pollutant in the box, $V.\Delta c$, must balance the mass gains from inflow and the emission source, having subtracted off the losses in mass from decay and outflow. That is to say:

$$V.\Delta c = S.\Delta t + aVPc_b.\Delta t - Vkc.\Delta t - aVc.\Delta t \quad (1)$$

dividing through by V and rearranging, we have:

$$dc/dt = S/V + aPc_b - (k+a)c \quad (2)$$

Assuming that the emission source strength and the background concentration remain constant, then eventually a steady state will be reached when dc/dt becomes zero. The steady state pollutant concentration, c_i, will then become:

$$c_i = aPc_b/(a+k) + S/(V(a+k)) \quad (3)$$

Equations (2) and (3) define time-dependent and steady state indoor air quality models, respectively (Dockery and Spengler,

1981; Hoek et al., 2013). We now provide some illustrative examples of the implementation of these models.

EXERCISE A1.1 DECORATING AN INDOOR SPACE

In this example, we envisage decorating an indoor space of 24 m^3 by applying decorative solvent-based paint at a rate of 1 litre per hour. The paint (or more correctly white spirit) has a density of 0.785 g/l (https://en.wikipedia.org/wiki/White_spirit) and a trace toxic component present at a level of 1 part in 1 million parts, 1 ppm, of white spirit solvent. We assume that the background and initial concentrations of the trace component are zero and that the indoor space is ventilated at a rate of 1 air change per hour. The trace component has no decay in the internal space and is lost only by ventilation.

The time history of the trace component, c, using equation (2) is given by:

$$dc/dt = S/V - ac \quad (4)$$

The concentration of the trace component comes into steady state at a concentration level given by:

$$ci = S/(Va) \quad (5)$$

which is 32.7 ng m^{-3}. With double the ventilation rate, the steady state concentration is lower and persists for a shorter period after painting ceases. With half the ventilation rate, the steady state concentration is higher and persists longer after cessation. Hence the instructions to use adequate ventilation whilst decorating indoor spaces.

A Python computer code (annexI_exerciseI1.pdf) is provided to show how the time-dependent indoor air quality model in equation (2) is solved numerically in this example.

EXERCISE A1.2 A POLLUTANT WITH BOTH INDOOR AND OUTDOOR SOURCES – ALBERT HALL

In this example, we answer the question about ventilation requirements for the Albert Hall in London, using the time-independent equation (3). The Albert Hall is a famous venue for music concerts and we will assume that it seats 5000 people and has an internal volume of 100,000 m^3 (estimates rounded from data in https://en.wikipedia.org/wiki/Royal_Albert_Hall). Each adult listening to the concerts is assumed to breathe in and out about 10,000 litres of air per day (United States Environmental Protection Agency, 2011). The exhaled breath of an adult at rest typically contains about 4% by volume carbon dioxide, (https://sciencing.com/functions-aveoli-lungs-7234440.html).

We need to estimate the ventilation rate that would be required to maintain the indoor carbon dioxide concentration below 1000 ppm, (https://kane.co.uk/knowledge-centre/what-are-safe-levels-of-co-and-co2-in-rooms/), the level at which the audience will start to feel stuffy and uncomfortable, assuming the Albert Hall is at full capacity. This ventilation air is assumed to be outside air which has a background carbon dioxide concentration, c_b, of 400 ppm during the mid-2010s, (https://gml.noaa.gov/ccgg/trends/global.html). Changing the above mixing ratios in ppm into concentrations in g/m^3, we note that 1000 ppm is 1.83 g m^{-3} and 400 ppm is 0.73 g m^{-3}, based an atmospheric density of 1.2041 kg m^{-3} at 20°C and the molecular mass of the atmosphere of 28.9644 g mol^{-1} (National Oceanic and Atmospheric Administration, 1976) and of CO_2 of 44 g mol^{-1}.

Each person breathes out 0.4/24 m^3 of carbon dioxide each hour. The gram-molecular volume of carbon dioxide is 22.4 litres per 44 grams (Kaye and Laby, 1966) so the mass emission rate is 32.7 g hr^{-1} per person. In steady state, the indoor carbon dioxide concentration is given by:

$$c_i = S/(Va) + c_b$$

If c_i is set at 1.83 g m^{-3} and c_b at 0.73 g m^{-3}, then a becomes just below 1.5 air changes per hour or a complete air change of the internal volume every 40 minutes.

A Python computer code (annexI_exerciseI2.pdf) is provided to show how the time-dependent indoor air quality model in equation (2) is solved numerically in this example.

EXERCISE A1.3 INDOOR PM LEVELS CLOSE TO A BUSY ROAD

In this example, we use the time-dependent indoor air quality model described in equation (2) to describe indoor PM levels in a house close to a busy road. This situation is introduced in Chapter 3 of the main body of this monograph.

For simplicity in this example, we will assume that the pollutant of concern is PM, particulate matter with a size less than 2.5 microns. We will assume that PM has a penetration factor, P, of 0.9, that is to say that one-tenth of PM particles are lost during ingress from the outside atmosphere. Assuming a decay time constant of 10 hours for the losses of PM inside the indoor environment, hence $k = 0.1\ hr^{-1}$. The appropriate version of the indoor air quality model is therefore:

$$V.\Delta c = S.\Delta t + aPVc_b.\Delta t - aVc.\Delta t - kVc.\Delta t \qquad (6)$$

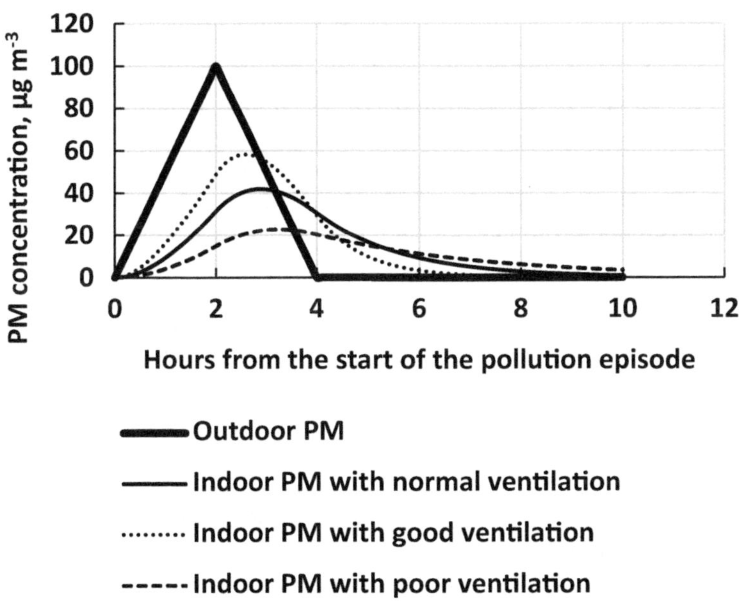

Figure A1.1 Diagram illustrating the relationship between indoor and outdoor air quality for PM using a time-dependent version of an indoor air quality model. The thick black line represents the time behaviour of the outdoor PM concentrations. The thin black line represents the numerical solution for the indoor PM concentration with normal ventilation, dotted with good ventilation and dashed with poor ventilation.

In this example, we envisage a pollution episode building up in the outdoor environment surrounding a house in an urban area in the vicinity of a busy road. Let us say that the pollution episode begins at 06:00 hours, reaches a peak at 08:00 hours and finishes at 10:00 hours. This is illustrated in Figure A1.1, where a solid black line represents the time behaviour of the outdoor PM concentration, c_b, which starts at zero, reaches 100 ug m^{-3} after two hours and declines back to zero after four hours.

We can solve equation (6) numerically by setting an incremental time step, Δt, of one-thousandth of an hour and start at time, $t = 0$, with the indoor PM concentration, c, at zero. We calculate the increment in the indoor PM concentration from the gain in PM from the inflow of background PM and the loss of PM in the outflow and decay. This increment is then added to the concentration at the start of the increment to obtain the concentration at the end of the time increment. This is then the new starting concentration for the next time increment and a new PM concentration increment is calculated. This process is then repeated until the endpoint of the calculation is reached.

A Python computer program (annexI_exerciseI3.pdf) is provided to illustrate how the time-dependent indoor air quality model is solved numerically in this example.

In Figure A1.1 we illustrate the numerical solution of the time-dependent indoor air quality model for the PM concentrations inside the home. The thick line shows an illustrative PM air pollution episode in an urban area. The peak in outdoor PM in this diagram shown as a thick solid line may, for example, represent the morning rush hour and extends over a four-hour period. The thin solid line represents the simple indoor air pollution model indoor PM concentrations that result from modest infiltration and ventilation rates. The magnitude of the indoor PM peak is less than half the outdoor PM peak, 42 µg m^{-3} instead of 100 µg m^{-3}, and is delayed by nearly an hour. If the ventilation rate is increased (dotted line), the magnitude of the PM peak increases by about 16 µg m^{-3} and it occurs about a half-hour sooner. However, the PM peak is dissipated more quickly as the home is better ventilated.

In contrast, decreasing the ventilation rate (dashed line) decreases the magnitude of the PM peak by 20 µg m^{-3} at the expense of delaying the timing of the maximum to an hour and a quarter after the outdoor peak and extending time period of the exposure to the PM.

Indoor air quality is thus strongly influenced by outdoor air quality, but the relationship is not straightforward, being dependent on the specific pollutant and ventilation rates.

REFERENCES

Dockery, D.W., Spengler, J.D., 1981. Indoor–outdoor relationships of respirable sulfates and particles. *Atmospheric Environment* 15, 335–343.

Hoek, G., Hanninen, O., Cyrys, J., 2013. *Indoor–outdoor relationships of particle number and mass in European cities*. In M. Viana (ed.), *Urban air quality in Europe*, pp. 321–337. Heidelberg, Germany: Springer.

Kaye, G.W.C.K., Laby, T.H., 1966. *Tables of physical and chemical constants*. London: Longman.

National Oceanic and Atmospheric Administration, 1976. *US standard atmosphere, 1976*. Report NOAA-S/T 76–1562. Washington, DC: National Oceanic and Atmospheric Administration.

United States Environmental Protection Agency, 2011. *Exposure factors handbook: 2011 edition*. EPA Report EPA/600/R-09/052F, Washington, DC: United States Environmental Protection Agency.

APPENDIX 2: URBAN AIR QUALITY MODELS

In this appendix, we assemble and run urban air quality models and apply them to some illustrative example situations. Initially, most urban air quality models were based on Gaussian plume models and so this is where we begin. More recently, with the increasing availability of computer resources, large three-dimensional Eulerian grid models are used for urban air quality modelling. These are basically regional air quality models where spatial resolution has been increased and model domains have shrunk to the urban scale. For further information on Eulerian grid models, the reader is referred to Appendix 3.

Gaussian plume models have been in use for decades to describe the concentrations found in the environment downwind of a single point source (Turner, 1970; Pasquill and Smith, 1983). They assume fixed geometrical shapes for the air pollutant concentration distribution based on the Gaussian or 'bell-shaped' curve. The dimensions of the geometrical shapes have been measured in field experiments under different weather conditions. Gaussian plume dispersion models have been used extensively across the globe to fix the heights of chimneys so as not to cause environmental damage downwind of large stationary sources of aggressive pollutants. Particular applications include setting chimney heights for large coal-burning power stations, smelters and oil refineries. This application of Gaussian plume models is illustrated in Exercise A2.1 below.

Early urban air quality models treated urban emissions as a set of hundreds to thousands of single sources of each air pollutant. The concentrations of the air pollutant at the receptor would be calculated as the sum of the concentration contributions from all the single sources in that urban area, at that point in time. The shapes of the Gaussian plumes would generally be taken from the literature. The wind speed and direction data, together with information on the atmospheric boundary layer depth would be taken from an off-line meteorological repository. Information on the instantaneous pollutant emission rates, chimney heights and effective stack heights would be taken from the emission inventory for that urban area, for that time period. The model would be advanced by the model time-step, typically one hour, and the calculation at each receptor would be repeated. In this way, a time history of the air pollutant concentration at the receptor would be built up.

Gaussian plume models, however, have some important drawbacks which have meant that they have lost favour over their competitors in recent years. It has proved difficult to represent removal processes for the atmospheric pollutants in Gaussian plume models. Dry and wet removal can be represented, in principle, using a simple first order loss coefficient such as Vg/h, where Vg is a dry deposition velocity and h is the atmospheric boundary layer depth. The loss of pollutant in travelling to the receptor is represented by exp(−Vgt/h) where t is the travel time. The effective emission rate employed is the actual emission rate, Q, multiplied by the depletion factor exp(−Vgt/h), that is Qexp(−Vgt/h). This source depletion method works well only for dry and wet deposition and radioactive decay where time constants are long and decay constants are fixed in space and are well understood.

However, for reactive pollutants such as NO_x, O_3 and PM, it has proved difficult to represent them accurately in Gaussian plume models. Here, the favoured approach is to increase the spatial resolution and decrease the size of the model domain in an appropriate regional air quality model. In many cases, this approach using Eulerian grid models has proved successful though it is not without difficulties. It has proved difficult to represent adequately the issues surrounding the effects of buildings on urban dispersion and street canyon effects. There are also difficulties with the availability of urban meteorological databases with sufficient accuracy and spatial resolution. These issues have not proved straightforward to address and remain very much a work in progress.

In Exercises A2.1 to A2.3 we show how Gaussian plume models can be applied in the urban context. We use these exercises to illustrate some features of urban air quality modelling without having to implement a complex Eulerian grid-based air quality model.

EXERCISE A2.1 AIRBORNE CONCENTRATIONS DOWNWIND OF A SINGLE POINT SOURCE OF POLLUTION

In this exercise, we calculate using a Gaussian plume model the airborne concentrations downwind of a single point source of pollution.

In a Gaussian plume model (Turner, 1970; Pasquill and Smith, 1983), the airborne concentration downwind of point source of pollution is given by:

$$C = Q/(2\pi\sigma_y\sigma_z u) \times \exp(-½(y/\sigma_y)^2) \times \{\exp(-½((z-h)/\sigma_z)^2 + \exp(-½(z+h)/\sigma_z)^2\} \quad (1)$$

where y is the distance of the receptor from the plume centre-line, z is the distance of the receptor above ground, h is the effective stack height, which is the sum of the physical chimney height and the plume rise, u is the wind speed at the receptor height, σ_y is the dispersion coefficient in the y direction, σ_z is the dispersion coefficient in the z direction and Q is the emission rate of the point source in g/s.

The dispersion parameters, σ_z and σ_y are functions of downwind distance, x, and are given by:

$$\sigma_z = Lx/(1 + (x/a))^q \quad (2)$$

and

$$\sigma_y = Kx/(1 + (x/a))^p \quad (3)$$

where K, L, a, p, q are parameters used to describe how the plume spreads vertically and horizontally with increasing downwind distance (Cota, 1984).

We have put together some input data for a Gaussian plume model in Table A2.1. The data are used to calculate the NO_x concentration at ground level on the centre-line of the plume at a downwind distance of 1 km from a large combined heat and power facility. If the imaginary source is emitting 1 tonne of NO_x as NO_2 s^{-1}, then the estimated NO_x concentration is 0.20 ug NO_2 m^{-3}. A Python program (annexII_exerciseII1.pdf) is provided to illustrate the calculation using equations (1) to (3) above. This program has been transcribed from a BASIC code from Cota (1984).

Table A2.1 Input data for the calculation of the concentration distribution downwind of a single point source of pollution using a Gaussian plume model.

Input parameter	*Parameter value*
x downwind distance in metres	1000
y distance from centre-line in metres	0
z height above ground in metres	0
Q emission rate of pollutant in kg NO_x as NO_2 s^{-1}	1000
u wind speed in m s^{-1}	6
L stability constant in km	47.5
a stability constant in km	0.707
q stability constant	0.465
K stability constant in km	78.7
p stability constant	0.135
H effective stack height metres	192.2
Downwind concentration in μg m^{-3}	0.20

EXERCISE A2.2 AIR POLLUTION CONCENTRATIONS GENERATED BY A BUSY ROAD

In this exercise, we use a Gaussian plume model to calculate the PM concentration distribution produced by motor vehicles travelling along a heavy-traffic road in a built-up area.

We begin with the simplest problem of this type, that is with a long straight road, located at ground level and passing through a relatively unobstructed urban area. Our focus is on the air pollutant concentrations produced by the motor vehicles on the road at short distances from it. At these short distances, the road can be treated as an infinite line source, Q, where Q is in g m^{-1} s^{-1}. Observations confirm that the pollutant distribution is largely insensitive to the angle of the road to the wind (Calder, 1973), so we will assume that the wind is perpendicular to the road.

The concentration of the pollutant, C, at a distance of x metres downwind of an infinite line source is given by:

$$C = \sqrt{2/\pi}\ Q/(ua(x + c)^b) \qquad (4)$$

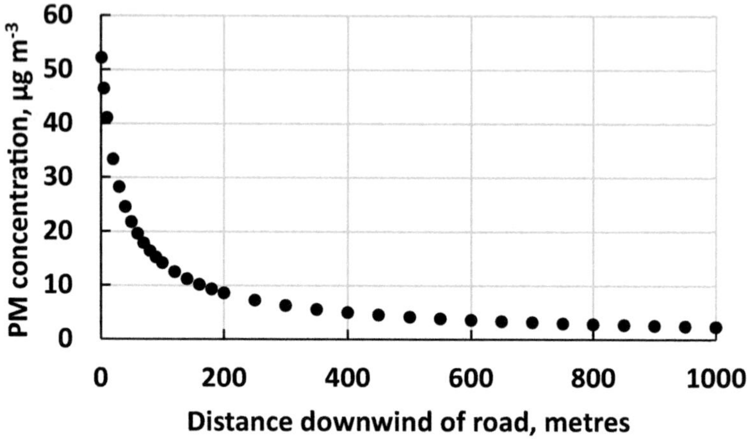

Figure A2.1 PM concentrations downwind from a busy road showing how concentrations fall off rapidly with distance.

where the stability constants a = 0.0856, b = 0.8650 and c = 27 metres, and the wind speed u = 1 ms^{-1} (Calder, 1973). For a road with a line source strength of 100 µg PM m^{-1} s^{-1}, equivalent to 36,000 vehicles per day emitting 10 mg PM per km travelled, the downwind PM concentrations are shown in Figure A2.1.

Concentrations of traffic pollutants therefore decline rapidly with distance away from busy roads, see Figure A2.1. Concentrations fall by one half roughly 30 metres from the road edge, by a further one quarter by 120 metres from the road edge and become undetectable at greater distances beyond 1 km.

A Python code (annexII_exerciseII2.pdf) is available illustrating the implementation of the infinite line source Gaussian plume model to reproduce the results in Figure A2.1.

EXERCISE A2.3 LEAKAGE OF NATURAL GAS IN THE LONDON AREA

In this exercise, we use observations of ethane, a trace component of natural gas to estimate the leakage of methane from the natural gas distribution system in London. Methane has a wide range of urban sources including domestic rubbish, waste disposal, humans and other animals, and it is not possible to gauge methane emissions from natural gas leakage based on methane measurements alone. Natural gas contains a trace component, ethane, which acts a tracer for natural gas leakage and allows us to work backwards from ethane observations to the natural gas leakage rate that is needed to support the ethane observations.

Natural gas leaks out of the distribution system that is used to provide natural gas for heating and cooking in homes throughout the London area. The distribution network is dense enough that it can be assumed to act like an area source of pollutants which is simply a two-dimensional network of point sources, each represented with a Gaussian plume model (Chamberlain et al., 1979; Hamilton and Mansfield, 1991). In this exercise, each point source acts as source of the tracer, ethane, with an area source strength of Q g m^{-2} s^{-1}. The ground level concentration, C in g m^{-3}, of ethane above this area source is given by:

$$C = \sqrt{(2/\pi)}\ Qr^{1-s} / ua(1 - s) \qquad (5)$$

where u is the wind speed which we take to be 5 m s^{-1}, a and s are dispersion parameters which we take as a = 0.05 km and s = 0.68, respectively, and r is the effective dispersion radius of London which is 8.55 km, based on Chamberlain et al., (1979); Hamilton and Mansfield, (1991). On this basis, we find C (in μg m^{-3}) = 20Q (in μg m^{-2} s^{-1}). A Python implementation of the area source Gaussian plume model is provided in annexII_exerciseII3.pdf.

Annual mean ethane (C_2H_6) levels at a central London monitoring station were 7 μg m^{-3} in 2011 (Derwent et al., 2017). Monitoring at Mace Head, Ireland indicated baseline levels of 1.6 μg m^{-3} (Derwent et al., 2012), which suggest a London elevation of 5.4 μg m^{-3}. Using the Gaussian plume area source model from above, this elevation must have been supported by an area source strength of 0.27 μg C_2H_6 m^{-2} s^{-1}. Taking a ratio by mass of 9.39 kg CH_4 per kg C_2H_6 for the composition of UK natural gas (MacCarthy, 2017), the estimated CH_4 source strength becomes 2.5 μg m^{-2} s^{-1} or about 80 tonne CH_4 km^{-2} yr^{-1}. This is much larger than the official published methane emission statistics for London (London Atmospheric Emissions Inventory, 2016), suggesting that official estimates significantly underestimate the leakage of natural gas.

REFERENCES

Calder, K.L., 1973. On estimating air pollution concentrations from a highway in an oblique wind. *Atmospheric Environment* 7, 863–868.

Chamberlain, A.C., Heard, M.J., Little, P., Whiffen, R.D., 1979. The dispersion of lead from motor exhausts. *Philosophical Transactions of the Royal Society A* 290, 577–589.

Cota, H.M., 1984. A Basic computer program for the Gaussian equation for a point source. *Journal of the Air Pollution Control Association* 31, 253.

Derwent, R.G., Field, R.A., Dumitrean, P., Murrells, T.P., Telling, S.P., 2017. Origins and trends in ethane and propane in the United Kingdom from 1993 to 2012. *Atmospheric Environment* 156, 15–23.

Derwent, R.G., Simmonds, P.G., O'Doherty, S., Grant, A., Young, D., Cooke, M.G., Manning, A.J., Utembe, S.R., Jenkin, M.E., Shallcross, D. E., 2012. Seasonal cycles in short-lived hydrocarbons in baseline air masses arriving at Mace Head, Ireland. *Atmospheric Environment* 62, 89–96.

Hamilton, R.S., Mansfield, T.A., 1991. Airborne particulate elemental carbon: its sources, transport and contribution to dark smoke and soiling. *Atmospheric Environment* 25A, 715–723.

London Atmospheric Emissions Inventory, 2016. *London atmospheric emissions inventory (LAEI) – methodology*. Stratford: Transport for London.

MacCarthy, J., 2017. *Gas composition data for the UK*, version 2. NAEI reference ED62328001/2015/AC9045/JMC. Didcot: Ricardo Energy & Environment.

Pasquill, F., Smith, F.B., 1983. *Atmospheric diffusion: Study of the dispersion of wind-borne material from industrial and other sources*. Edited by R.S. Scorer. Ellis Horwood Series in Environmental Sciences. ISBN 0-1305-13350: Prentice-Hall Europe.

Turner, D.B., 1970. *Workbook of atmospheric dispersion estimates*. Publication no. AP-26. Durham, NC: Office of Air Programs, US Environmental Protection Agency.

APPENDIX 3: REGIONAL AIR QUALITY MODELS

In this appendix, we describe how to model regional air quality and how to apply regional air quality models in some illustrative situations.

Regional air quality models aim to relate the emissions of air pollutants to their environmental effects on the regional scale with the aim of aiding policy support and development. They assemble understanding of emissions, atmospheric transport, transformation and environmental effects into a single modelling system.

The policy framework underpinning our understanding of regional air pollution is illustrated in Figure A3.1. This framework is at the heart of most regional air quality models used for policy development.

A policy-maker would begin at the right-hand side of the framework diagram with the environmental damage (receptors) column. Policy-makers express commitments to reduce damage in these receptor systems within their own jurisdictions. Environmental scientists identify the causative agents that are causing the damage and these are listed in the third (effects) column of the framework, with lines showing the linkages between the causative agents and the environmental damage. Environmental acidification links to lakes, ground-waters and soils because this is the mechanism driving the loss of fishery status. The over-supply of nutrients (eutrophication) links to the loss of biological diversity in soils and plant ecosystems and to the formation of algal blooms in shallow seas such as the Baltic Sea and in lakes. Fine particles ($PM_{2.5}$) damage human health and contribute to global climate change. Elevated regional O_3 levels lead to exceedance of air quality standards set for the protection of human health and for the prevention

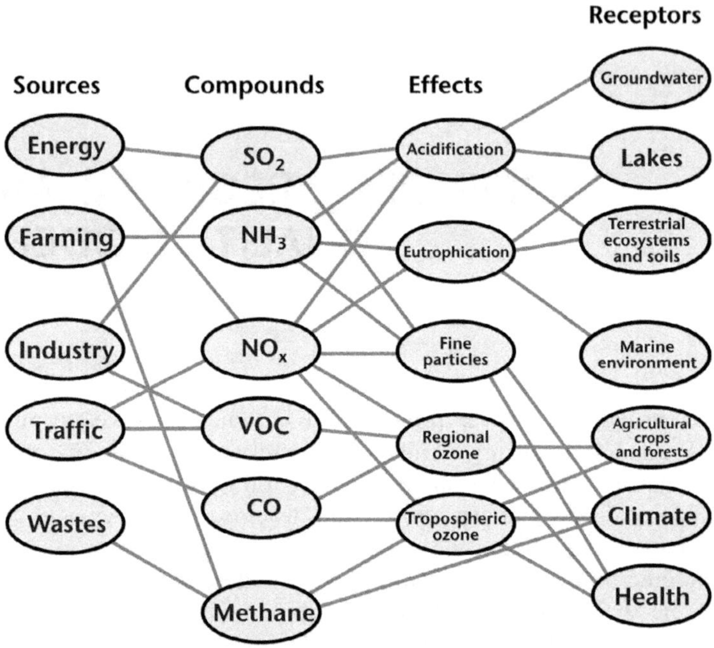

Figure A3.1 The multi-pollutant multi-effect framework underpinning our understanding of regional air pollution problems.
Source: redrawn from Grennfelt et al. (1994), by permission of Springer Nature.

of damage to crops and vegetation. There is also an important synergy between regional O_3 and global climate change.

The linkages between the second (compounds) and third (effects) columns are based on atmospheric science because these linkages record the atmospheric transformations that convert primary pollutants in the second column to the causative agents or secondary pollutants in the third column. Thus, the primary pollutants SO_2, NO_x and NH_3 all take part in the chemical transformations that drive acid rain, eutrophication and fine particle formation. VOCs and NO_x drive ground-level ozone formation as shown by the linkages between the second and third columns.

The column (sources) on the left-hand side of the framework represents the sources of the emissions of the primary pollutants.

This is where the policy-makers can exert their influence if they want to reduce environmental damage on the right-hand side (receptors) of the framework. The framework identifies a wide range of emission sources involved with regional air pollution. These are not just the large electricity generating power stations but motor vehicle traffic, agriculture and industry.

In the sections which follow, we discuss emissions data, atmospheric dispersion and transport, transformations and environmental effects and how they are described within the mathematical framework of a regional air quality model. We finish with some illustrative exercises which address the sulphur and oxidised nitrogen budgets for the United Kingdom and ground-level ozone formation across north-west Europe.

EMISSIONS DATA AND EMISSION INVENTORIES

An emission inventory is a list of pollutant sources within a given region which tabulates the emissions of that pollutant within a given timescale. The inventory provides the main mechanism for the presentation and implementation of emissions data for policy development and regional air quality modelling. Regional emission inventories typically cover specific cities, industrial centres, counties, states or countries and address SO_2, NO_x, CO, VOCs, CH_4 and PM. Emission inventories are continually improving in accuracy, spatial and temporal resolution, coverage of VOCs and PM size ranges.

Emissions are estimated using the relationship: emission = activity × emission factor, where activity is some statistical quantity such as vehicle kilometres driven, tonnes of fuel burned or mass of product manufactured and the emission factor is the mass of pollutant released per unit of activity. There are various sources of statistical data with which to estimate activities. These include energy statistics, traffic surveys and traffic counts, housing stock data and transport statistics. Emission factors are specific to each pollutant and to each industrial process, equipment, vehicle and building, for example. They depend on fuel, operating conditions, time of day, day of the week and season. For large pollution sources such as power stations, industrial complexes and oil refineries, emissions can be measured in real time. Vehicle emissions can be measured on a rolling-road (chassis dynamometer) in a test-bed facility or under real driving conditions with on-board vehicle emission monitors.

METEOROLOGICAL DATA

Most pollutant emissions are emitted into the lowest layers of the atmosphere, close to the ground surface, into the atmospheric boundary layer. Meteorological data are required to describe how pollutants are dispersed in the boundary layer and are transported by the wind away from their sources and into the downwind environment towards the receptors. Calm conditions are relatively rare and boundary layer turbulence efficiently disperses pollutants on time scales of minutes to hours. Wind speed and wind direction data are used to describe the path taken by pollutants over periods from hours through to days. Temperature and relative humidity data are used to fix the rates of chemical reactions which describe the production and loss of chemically reactive pollutants. Precipitation data are used to quantify the rates of removal of water-soluble pollutants.

Historically, the meteorological data employed in regional air quality models were assembled from the available weather records or from weather forecasts and were processed separately outside of the regional air quality model. The meteorological data were said to have been processed off-line from the regional air quality model. Over the years, computer capacity has increased phenomenally and now it has become feasible to build the regional air quality model inside the framework of the numerical weather predictions models used to forecast the weather. In this way, there are no problems about conflicting grid meshes, model domains and the storage of large meteorological databases. Meteorological data are now processed on-line. This has allowed the air quality predictions to influence the meteorological development through a chemical feedback between particles, clouds and radiation. On-line models, in principle, offer more accurate estimates to be made of boundary layer depths, wet removal of water-soluble species, cloud amounts and natural biogenic sources of organic compounds.

ATMOSPHERIC TRANSFORMATIONS

Not all harmful regional air pollutants are emitted directly into the atmosphere, (primary pollutants), and some are formed in the atmosphere by atmospheric chemical processes or transformations involving the emitted species. The pollutants formed in the atmosphere are called secondary pollutants and the pollutants from which they are formed are termed precursors. Some secondary pollutants may have additional primary sources.

The oxides of nitrogen, $NO_x = NO + NO_2$, are a case in point. Nitric oxide (NO) is wholly primary in origins whereas nitrogen dioxide (NO_2) has both primary emissions and is formed in the atmosphere by reaction of NO with ozone (O_3):

$$NO + O_3 \rightarrow NO_2 + O_2$$

This is an example of a simple transformation reaction.

Acid deposition is driven by the dry and wet deposition of sulphur species, of which sulphur dioxide (SO_2) has primary origins whereas sulphuric acid (H_2SO_4) and ammonium sulphate ($(NH_4)_2SO_4$) have secondary origins. It is also driven by the dry and wet deposition of oxidised nitrogen species, of which NO_2 is both primary and secondary in origins, and nitric acid (HNO_3) and ammonium nitrate (NH_4NO_3) are wholly secondary.

Ground-level ozone is entirely of secondary origins since there are no significant direct emissions of ozone into the atmosphere. Almost all of the ozone found in the lower atmosphere was formed there by the reactions of organic compounds with the oxides of nitrogen in the presence of sunlight. There may be upwards of thousands of emitted organic compounds from man-made and natural biogenic sources that participate in ground-level ozone formation. Consequently, considerable assumptions, simplifications and approximations must be made in the representation of ground-level ozone formation in regional air quality models.

Understanding these transformations and representing them accurately in regional air quality models is an important and essential prerequisite. Information on the chemical pathways and transformations is provided within the regional air quality model in the form of a chemical mechanism. Although computer resources have grown dramatically in recent years, much of this increase in resources has been taken up by handling meteorological data on-line, by increasing the spatial resolution and by increasing the size of the spatial domain treated. Hence most regional air quality models can handle only cut-down and highly parameterised chemical species and chemical mechanisms, representing at best only several hundred chemical reactions. These limitations have exerted an influence on the ability of regional air quality models to represent accurately ground-level ozone and fine particle formation.

ENVIRONMENTAL IMPACTS

If a policymaker is to apply a regional air quality model to design policies to reduce environmental damage or to evaluate the likely impact of a given policy, then the regional air quality model needs to have a means of quantifying environmental impacts. As Figure A3.1 illustrates, each environmental receptor has its own environmental effect. A regional air quality model will need to be able to quantify the effects of acidification, eutrophication, fine particles on human health, and ground-level ozone on human health, crops and vegetation, as appropriate.

Environmental research programs have been set up with the explicit aim of quantifying environmental impacts in each of the receptors in Figure A3.1. In general, the severity of the environmental impacts increases with increasing exposure to damaging pollutants. Research programs have attempted to identify the damaging pollutant driving each environmental impact, and the so-called dose-response relationship that describes how particular impacts increase in severity with increasing exposure to the damaging agents. Each environmental receptor in Figure A3.1 has its own causative agent and dose-response relationship.

Lakes, streams and ground-waters have been acidified, particularly in Scandinavia, with the resulting loss of fishery status and of biological diversity. Extensive research programmes have identified soil acidification as the root cause of the environmental damage and this has, in turn, been driven by the deposition of acidic sulphur and oxidised nitrogen species on sensitive upland catchments, where the soil mineralogy implies limited buffering capacity from calcareous minerals. It has been possible to define a critical deposition load that a catchment can bear without long-term damage occurring. When the deposition load is above the critical load, acidification damage occurs. Using a regional air quality model, it is possible to overlay a deposition map with a critical loads map and estimate the ecosystem and catchment areas where critical loads are exceeded, and to test whether a given policy reduces critical loads exceedance.

The critical loads approach has also been employed for eutrophication damage and the loss of biological diversity in sensitive plant and vegetation ecosystems that it drives. With this receptor, research programmes have identified that the causative agents are the dry and wet deposition of the reduced (NH_3, NH_4NO_3 and $(NH_4)_2SO_4$) and oxidised (NO_2, HNO_3 and NH_4NO_3) forms of nitrogen (see Figure A3.1), driven largely by local ammonia (NH_3) emissions from intensive agriculture.

For agricultural crops and forests, a different approach to the quantification of ozone damage is appropriate because, here, concern is with ozone pollution episodes and not long-term deposition as with environmental acidification and eutrophication. Through extensive research programmes, it has been possible to identify a threshold for ozone damage to crops and forests at levels of the order of 40 ppb as an hourly mean level. A quantity called the Accumulated Ozone above this Threshold AOT_{40} is then calculated by subtracting the threshold from each hourly mean level and summing over the entire growing period. Again, policies can be assessed in terms of their ability to reduce AOT_{40} levels.

For the receptor human health, identification of causative agents and dose-response relationships generally have involved experimental studies involving healthy volunteers or epidemiological studies involving whole urban populations or sensitive population groups. This quantification has led ultimately to the derivation of no-effects thresholds, air quality standards, air quality guidelines, after considerable detailed review and assessment of the available peer-reviewed literature. It is usual to assess policies for their impact on the non-attainment of air quality standards and guidelines.

MATHEMATICAL FRAMEWORKS

Regional air quality models may be classified according to their mathematical frameworks as either Lagrangian trajectory models or Eulerian grid models. In a Lagrangian trajectory model, a coordinate system is employed that moves with the winds carrying the pollutants. In a Eulerian grid model, the coordinate system is fixed in space, with the winds carrying the pollutants blowing through it.

These two frameworks are illustrated in Figures A3.2 and A3.3, respectively. A common feature of each approach is a model domain that includes the both the source and receptor regions, a gridded emission inventory for each of the relevant air pollutants and a meteorological database containing three-dimensional wind fields.

In a Lagrangian model, Figure A3.2, the emission grid acts as a constant and continuous source of air parcels. Each air parcel contains the appropriate mass of each pollutant, depending on the location within the emission grid and the time of day etc. At the end of each model time step, Δt, each air parcel is moved to its new location depending on its location within the three-dimensional wind-field at that time. If an air parcel moves out of the model domain then it is deleted from the model run. The number of air parcels thus builds up with model run time and there may be upwards of millions of air parcels in a long-range transboundary model run. The mass of each pollutant remaining in each air parcel is calculated by subtracting the masses lost and adding the masses gained during the time step to the masses at the start of the time step. Mass is gained or lost by chemical transformations, lost by deposition to the underlying surface and gained or lost by exchange with neighbouring air parcels.

Pollutant concentrations at a receptor in Figure A3.2 are calculated by constructing an imaginary box around the receptor at any point in time and ascertaining which of the air parcels are within this box. The concentration of each pollutant at the receptor is then obtained by

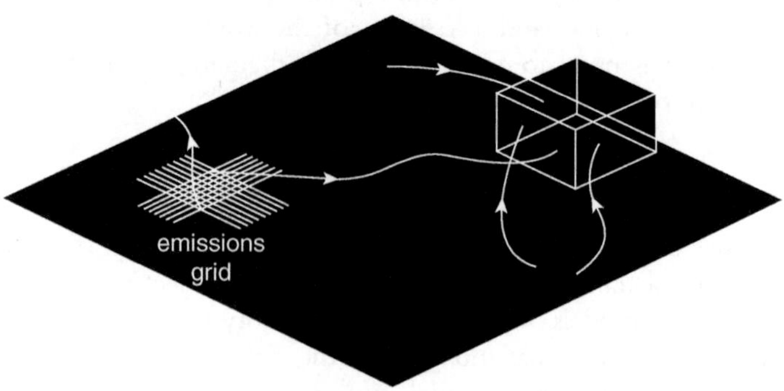

Figure A3.2 Diagrammatic representation of a Lagrangian trajectory model.

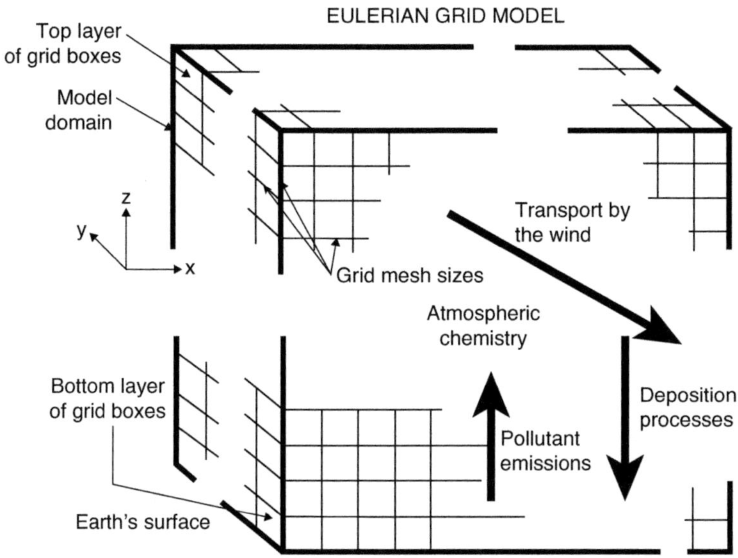

Figure A3.3 Diagrammatic representation of a Eulerian grid model.

summing the masses of each pollutant carried by the air parcels within the box and dividing by the box volume. If an air parcel does not remain within the model domain, then, in principle, it could be deleted from the model run, since it takes no part in fixing the output concentrations. Because of this simplification, Lagrangian models operate by handling the meteorological data in an off-line mode. All pollutants are advected together in one step and this represents a huge saving in computer resources, especially if detailed chemical mechanisms are employed to represent the transformations. However, these days highly simplified and parameterised chemical mechanisms remain the norm and on-line meteorological models are preferred. Consequently, Lagrangian trajectory models have lost ground significantly to Eulerian grid models.

In a Eulerian model, a three-dimensional grid mesh is erected over the emissions grid with its lower boundary at Earth's surface and upper boundary at the tropopause (see Figure A3.3). Depending on the grid mesh spacing, there may be upwards of a million grid cells in the model domain. In each model time step, the three-dimensional wind field is resolved along the axes of the grid and each wind component is employed to move pollutants from one grid cell to the next nearest neighbours at a rate proportional to the resolved wind components. Pollutants are added to each box from the emissions grid if the grid box has its lower boundary on Earth's surface. Pollutants are added and lost by chemical transformation, lost to Earth's surface by deposition and gained or lost by exchange with nearest neighbours. This calculation is repeated for each advected pollutant and for the entire set of grid boxes. Computer requirements are therefore huge for Eulerian air quality models with high spatial resolution. However, if these resources are available, modellers are rewarded with output results covering the entire model domain in three-dimensions at each model step. This may offer a huge policy advantage if large geographical areas are of concern or when it is important to identify pollution and deposition hot-spots or where air quality monitoring is sparse.

EXERCISE A3.1 SULPHUR BUDGET FOR THE UNITED KINGDOM

In this exercise, a box model is employed to quantify the sulphur deposition budget for the United Kingdom.

We begin by assembling a well-mixed box model over the United Kingdom. The box is ventilated by clean air from the North Atlantic Ocean passing from west to east, with polluted air moving out over the North Sea into continental Europe. We assume a crosswind orientation of north-south and box dimension, y, of 1000 km. We take the downwind box-dimension, x, to be 600 km and its orientation to be west-east. The height of the box, h, is set at 800 metres, the average height of the atmospheric boundary layer over the UK (Smith and Hunt, 1978). The mean windspeed, v, ventilating the box is taken to be 10.4 m s^{-1} (Smith and Hunt, 1978).

The UK SO_2 emission source strength, E, for the year 2010 can be taken as 460 thousand tonnes SO_2 per year (NAEI, 2021). Using a molar mass for SO_2 as 64.0 g mol^{-1} and 32.0 g mol^{-1} for S and 3.1536×10^7 seconds per year, we find E = 7293 g S s^{-1}. Sulphur dioxide takes part in various transformations which are represented in Figure A3.4 leading to the dry deposition of SO_2

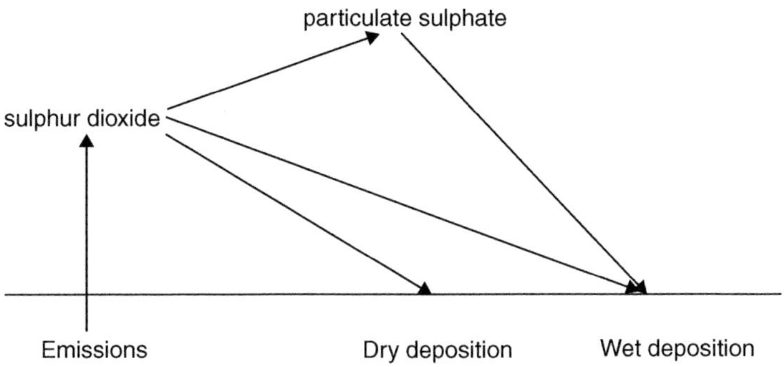

Figure A3.4 Illustrative diagram of the sulphur budget processes.

and the wet deposition of SO_2 and SO_4 aerosol. The aim is to estimate the total deposition fluxes of SO_2 and SO_4 aerosol to the UK surface area using the box model.

We first build the budgets for SO_2 using the box model. In a time interval, Δt, the gain in mass of SO_2 is given by $xyh\Delta c_{SO2}$, where c_{SO2} is the SO_2 concentration in mass of S per unit volume. This gain in mass is provided by emissions, having subtracted the loss to dry and wet deposition of SO_2, conversion of SO_2 to SO_4 and the loss of SO_2 by transport out of the model domain by the wind.

The gain in SO_2 mass from emissions is given by $E\Delta t$. We calculate the loss of SO_2 to dry deposition using the deposition velocity concept. The dry deposition velocity, Vg, of a pollutant is defined as the flux of pollutant per unit area and unit time, D, divided by the mass concentration of the pollutant, c, that is: $Vg = D/c$. For SO_2, field experiments have characterised an average deposition velocity of SO_2 as 5 mm s^{-1} for the UK surface (Metcalfe et al., 1995). Particulate sulphate (SO_4) is not readily dry deposited and has been neglected in the present exercise and so has been omitted from Figure A3.4. The mass of SO_2 as S deposited by dry deposition in the time interval, Δt, is thus $xyD\Delta t$ or $xyVgc_{SO2}\Delta t$.

Estimating the mass of SO_2 removed in wet deposition by rainfall is not straightforward because rainfall scavenging is a sporadic and intermittent process. In northwest Europe, rainfall scavenging can be described as an intermittent process occurring during wet periods, with the average length of wet periods being about 8 hours compared to 40 hours for the intervening dry periods. Rainfall scavenging is an efficient process for aerosol particles during rain, but is much less efficient for gases. Indeed, the global mean timescale for the removal of the most soluble of all gaseous compounds, water vapour, during rainfall is ten days or thereabouts. We construct a first order loss rate coefficient, Λ, a so-called constant drizzle loss coefficient that averages over the wet and dry periods, and the steady removal of water-soluble species during the wet periods. For particulate SO_4, we set $\Lambda_{SO4} = 1.3 \times 10^{-5}\ s^{-1}$ and for SO_2 $\Lambda_{SO2} = 1.0 \times 10^{-6}\ s^{-1}$ (Metcalfe et al., 1995). The mass of SO_2 as S deposited in wet deposition is thus $xyhc_{SO2}\Lambda_{SO2}\Delta t$.

We now estimate the loss of mass transported out of the model box by the advection wind, v. Since this occurs through the whole box face, the loss of mass is $yhvc_{SO2}\Delta t$. If the transformation of SO_2 to SO_4 can be described by a first order transformation coefficient, k s^{-1}, then the loss of SO_2 mass by transformation is $xyhkc_{SO2}\Delta t$. We take $k = 2.8 \times 10^{-6}\ s^{-1}$, that is 1% per hour, representing the oxidation of SO_2 in cloud droplets by hydrogen peroxide, H_2O_2 (Metcalfe et al., 1989). We neglect advection of sulphur into the box model.

Building a SO_2 mass budget for our box we find:

$$xyh\Delta c_{SO2} = E\Delta t - xyVgc_{SO2}\Delta t - xyh\Lambda_{SO2}c_{SO2}\Delta t - yhvc_{SO2}\Delta t - xyhkc_{SO2}\Delta t \quad (1)$$

Simplifying:

$$dc_{SO2}/dt + c_{SO2}(Vg/h + \Lambda_{SO2} + v/x + k) = E/xyh \quad (2)$$

which in steady state, when dc/dt = 0, becomes:

$$c_{SO2} = E/(xyh(Vg/h + \Lambda_{SO2} + v/x + k)) \quad (3)$$

Substituting the parameter values appropriate to SO_2, we find:

$$(Vg/h + \Lambda_{SO2} + v/x + k) = 2.7 \times 10^{-5}\ s^{-1} \quad (4)$$

giving the SO_2 concentration as 0.55 μg S m^{-3}, which is a reasonable estimate based on observations for 2010 (https://uk-air.defra.gov.uk/data/data-selector_service).

Since the oxidation of SO_2 is the sole source of the particulate SO_4, we can put together the atmospheric budget for particulate SO_4 as follows:

$$xyh\Delta c_{SO4} = xyhkc_{SO2}\Delta t - xyh\Lambda_{SO4}c_{SO4}\Delta t - yhvc_{so4}\Delta t \quad (5)$$

Simplifying:

$$dc_{SO4}/dt + c_{SO4}(\Lambda_{SO4} + v/x) = kc_{SO2} \quad (6)$$

In steady state then, $dc_{SO4}/dt = 0$ and

$$c_{SO4} = kc_{SO2}/(\Lambda_{SO4} + v/x) \quad (7)$$

giving the particulate SO_4 concentration as 0.051 μg S m^{-3}.

The Python code implementing the box model for the sulphur budget over the United Kingdom is available as annexIII_exerciseIII1.pdf.

The total sulphur deposition flux is thus given by the sum of the dry and wet deposition of SO_2 and the wet deposition of SO_4, as follows: $xyVgc_{SO2}\Delta t + xyh\Lambda_{SO2}c_{SO2}\Delta t + xyh\Lambda_{SO4}c_{SO4}\Delta t$, which may be simplified as: $xy(Vgc_{SO2} + h\Lambda_{SO2}c_{SO2} + h\Lambda_{SO4}c_{SO4})\Delta t$. If we set Δt as one year, 3.1536×10^7 seconds, and substitute the above parameter values, the total sulphur deposition flux becomes 71 thousand tonnes S yr^{-1}, or about 30% of UK emissions. This estimate is close to that obtained with more detailed regional air quality models for the selected region of the UK for 2010.

EXERCISE A3.2 OXIDISED NITROGEN BUDGET FOR THE UNITED KINGDOM

In this exercise, a box model is employed to quantify the oxidised nitrogen deposition budget for the United Kingdom.

We begin by assembling the same box model over the United Kingdom as we employed for Exercise A3.1 above. Figure A3.5 presents an illustrative diagram of the budget processes that we envisage controlling the oxidised nitrogen budget over the United Kingdom. The processes are initiated by the emission of NO_x = NO + NO_2 which occur largely as nitric oxide (NO). NO is rapidly converted into nitrogen dioxide (NO_2) on the timescales of

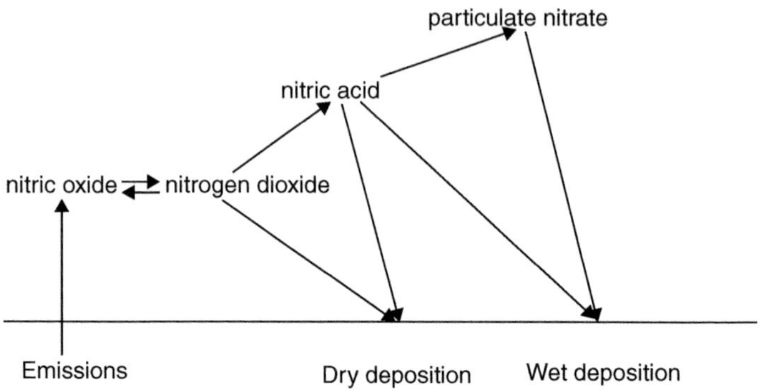

Figure A3.5 Illustrative diagram of the oxidised nitrogen budget processes.

minutes, as shown in Figure A3.5. NO_2 is subsequently oxidised to gaseous nitric acid (HNO_3) by hydroxyl (OH) radicals or removed by dry deposition to Earth's surface. HNO_3 is in turn converted to particulate nitrate by reaction with sea-salt or ammonia or removed by dry deposition and wet deposition. Particulate nitrate (NO_3) is largely removed by wet deposition or, like NO_2 and HNO_3, transported out of the model box by the wind.

We first build the budgets for NO_2 using the box model. In time interval, Δt, the gain in mass of NO_2 is given by $xyh\Delta c_{NO2}$, where c_{NO2} is the NO_2 concentration in mass of N per unit volume units. This gain in mass is provided by emissions, having subtracted the loss to dry deposition of NO_2, conversion of NO_2 to HNO_3 and the loss of NO_2 by transport out of the model domain by the wind.

The gain in NO_2 mass from emissions is given by $E\Delta t$. We calculate the loss of NO_2 to dry deposition using the deposition velocity concept. For NO_2, field experiments have characterised an average deposition velocity, Vg_{NO2}, of NO_2 as 1.5 mm s^{-1} for the UK surface. The mass of NO_2 as N deposited by dry deposition in the time interval, Δt, is thus $xyVg_{NO2}c_{NO2}\Delta t$. NO_2 is not sufficiently soluble in water for wet deposition to be efficient and it is neglected in Figure A3.5.

We now estimate the loss of mass transported out of the model box by the advection wind, v. Since this occurs through the whole box face, the loss of mass is $yhvc_{NO2}\Delta t$. Assuming that the transformation of NO_2 to HNO_3 can be described by a first order transformation coefficient, k_{NO2} s^{-1}, then the loss of NO_2 mass by transformation is $xyhk_{NO2}c_{nO2}\Delta t$. We take $k_{NO2} = 4.3 \times 10^{-5}$ s^{-1}, that is 15% per hour (Metcalfe et al., 1989), representing the rapid gas-phase oxidation of NO_2 by hydroxyl (OH) radicals and by ozone.

Building a NO_2 mass budget for our box we find:

$$xyh\Delta c_{NO2} = E\Delta t - xyVg_{NO2}c_{NO2}\Delta t - yhvc_{NO2}\Delta t - xyhk_{NO2}c_{NO2}\Delta t \quad (8)$$

Simplifying and putting dc_{NO2}/dt in steady state gives:

$$C_{NO2} = E/(xyh(Vg_{NO2}/h + v/x + k_{NO2})) \quad (9)$$

The UK NO_x (= NO_2) emission source strength, E, for the year 2010 can be taken as 1230 thousand tonnes NO_2 per year (NAEI, 2021). Using a molar mass for NO_2 as 46.0 and 14.0 for N and 3.1536×10^7 seconds per year, we find E = 11870 g N s^{-1}. Setting the above parameters for NO_2, the steady state NO_2 concentration becomes 0.40 μg N m^{-3}, which is a reasonable estimate based on annual mean rural monitoring network observations for 2010 (https://uk-air.defra.gov.uk/data/data-selector_service).

Since the oxidation of NO_2 is the sole source of the HNO_3, we can put together its atmospheric budget as follows:

$$xyh\Delta c_{HNO3} = xyhk_{NO2}c_{NO2}\Delta t - xyVg_{HNO3}c_{HNO3}\Delta t - xyh\Lambda_{HNO3}c_{HNO3}\Delta t - yhvc_{HNO3}\Delta t - xyhk_{HNO3}c_{HNO3}\Delta t \quad (10)$$

where Vg_{HNO3} is the dry deposition velocity for HNO_3 set at 40 mm s^{-1} (Metcalfe et al., 1989) since it is very readily taken up by Earth's surface, Λ_{HNO3} is the constant drizzle wet scavenging coefficient set as 9.0×10^{-6} s^{-1} (Metcalfe et al., 1989) because it is a highly soluble gas and k_{HNO3} is the first order transformation coefficient for the reaction of nitric acid with ammonia (NH_3) set as 2.4×10^{-4} s^{-1} because it is a rapid reaction (Metcalfe et al., 1995).

In steady state, then, $dc_{HNO3}/dt = 0$ and

$$C_{HNO3} = k_{NO2}c_{NO2}/(\Lambda_{HNO3} + Vg_{HNO3}/h + v/x + k_{HNO3}) \quad (11)$$

giving the HNO_3 concentration as 0.054 μg N m^{-3}.

Since the reaction of HNO_3 is the main source of the particulate nitrate (NO_3), we can put together its atmospheric budget as follows:

$$xyh\Delta c_{NO3} = xyhk_{HNO3}c_{HNO3}\Delta t - xyh\Lambda_{NO3}c_{NO3}\Delta t - yhvc_{NO3}\Delta t \quad (12)$$

where Λ_{NO3} is the constant drizzle wet scavenging coefficient set as 1.3×10^{-5} s^{-1} (Metcalfe et al., 1989) because it is a highly soluble particulate species.

In steady state then, $dc_{NO3}/dt = 0$ and

$$C_{NO3} = k_{HNO3}c_{HNO3}/(\Lambda_{NO3} + v/x) \quad (13)$$

giving the particulate NO_3 concentration as 0.43 μg N m^{-3}.

The Python code implementing the box model for the oxidised nitrogen budget over the United Kingdom is available as annex-III_exerciseIII2.pdf.

The total oxidised nitrogen deposition flux is thus given by the sum of the dry deposition of NO_2, the dry and wet deposition of HNO_3 and the wet deposition of NO_3, as follows: $xyVg_{NO2}c_{NO2}\Delta t + xyVg_{HNO3}c_{HNO3}\Delta t + xyh\Lambda_{HNO3}c_{HNO3}\Delta t + xyh\Lambda_{NO3}c_{NO3}\Delta t$, which may be simplified as: $xy(Vg_{NO2}c_{NO2} + Vg_{HNO3}c_{HNO3} + h\Lambda_{HNO3}c_{HNO3} + h\Lambda_{NO3}c_{NO3})\Delta t$. If we set Δt as one year, 3.1536×10^7 seconds, and substitute the above parameter values, the total oxidised nitrogen deposition flux becomes 144 thousand tonnes N yr^{-1}, or about 40% of UK NO_x emissions. This percentage is larger than the corresponding value for sulphur found in Exercise A3.1 because of the large dry deposition velocity for HNO_3 compared with SO_2.

EXERCISE A3.3 REGIONAL PHOTOCHEMICAL OZONE FORMATION

In this exercise, we consider the role played by organic compounds in regional-scale ground-level ozone (O_3) formation. We examine the transformations that take place leading to elevated O_3 levels from the oxidation of organic compounds in the presence of the oxides of nitrogen and sunlight. We quantify the photochemical O_3 production from organic compounds emitted from man-made

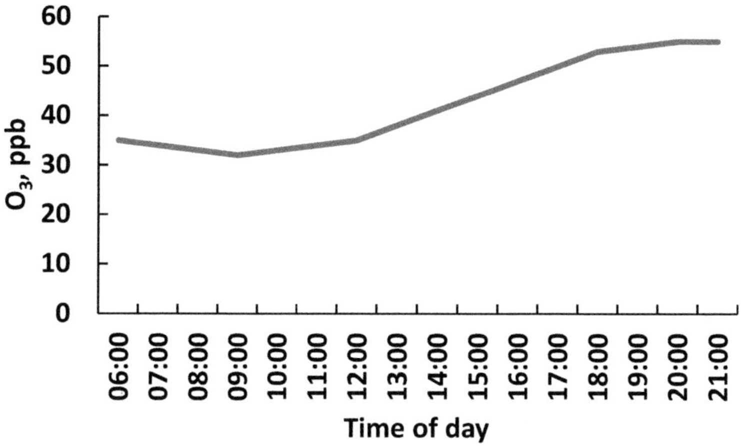

Figure A3.6 Behaviour of ozone in an air parcel traversing an urban or industrial area.

sources under conditions appropriate to north-west Europe and show how it amounts to several ppb per hour.

We begin by considering an air parcel traversing north-west Europe, see Figure A3.6, and initially containing O_3 at baseline levels in the range from 30–50 ppb. As it enters an urban or industrial region, O_3 levels drop as NO_x levels rise and O_3 is converted into NO_2. This chemical reaction leaves oxidants, $O_x = O_3 + NO_2$ conserved as it merely converts one oxidant into another. As the air parcel traverses the urban or industrial region it picks up emissions of organic compounds in addition to NO_x. Eventually the air parcel moves into the downwind environment and NO_x levels start to fall. The oxidant balance shifts back in favour of O_3 with NO_2 levels decreasing, as shown in Figure A3.6.

If the emissions of organic compounds have been large enough, photochemical ozone production will also begin as NO_x levels fall and O_3 levels will start to rise above baseline levels as shown in Figure A3.6, reaching several ppb per hour. Photochemical O_3 production is driven by the oxidation of organic compounds following their reaction with hydroxyl (OH) radicals. We can illustrate this process by following the reaction of ethylene (ethene, C_2H_4) with OH:

$$OH + CH_2 = CH_2 \rightarrow OHCH_2CH_2 \quad (14)$$

$$OHCH_2CH_2 + O_2 \rightarrow OHCH_2CH_2O_2 \quad (15)$$

$$OHCH_2CH_2O_2 + NO \rightarrow OHCH_2CH_2O + NO_2 \quad (16)$$

$$NO_2 + h\nu \text{ (sunlight)} \rightarrow NO + O \quad (17)$$

$$O + O_2 \rightarrow O_3 \quad (18)$$

$$OHCH_2CH_2O + O_2 \rightarrow HCHO + HCHO + HO_2 \quad (19)$$

$$HO_2 + NO \rightarrow OH + NO_2 \quad (20)$$

$$NO_2 + h\nu \rightarrow NO + O \quad (21)$$

$$O + O_2 \rightarrow O_3 \quad (22)$$

Following the attack of OH on the ethylene molecule, a series of organic radicals ($OHCH_2CH_2$), organic peroxy radicals ($OHCH_2CH_2O_2$), alkoxy radicals ($OHCH_2CH_2O$) and carbonyl compounds (HCHO) are formed in a series of rapid consecutive reactions. The OH radical is recycled through the hydroperoxyl (HO_2) radical. These reactions lead to the production of two O_3 molecules for every ethylene molecule reacting. Each ethylene molecule is converted into two molecules of formaldehyde (HCHO) which is itself highly reactive and may go on to form more molecules of O_3. Almost all man-made and biogenic organic compounds follow a similar reaction pathway to that for ethylene after their initial reaction with OH radicals.

The concentrations of an organic compound, [OC] and $[O_3]$ in an air parcel leaving the urban and industrial region can be represented by the following equations, assuming that the air parcel is a well-mixed box:

$$d/dt[OC] = -k[OH][OC] \quad (23)$$

$$d/dt[O_3] = \alpha k[OH][OC] - [O_3]Vg/h \quad (24)$$

where k is the rate coefficient for the reaction of OH with the organic compound, α molecules of O_3 are produced for every molecule of organic compound reacted, Vg is the dry deposition velocity for O_3 and h is the height of the air parcel, typically the depth of the atmospheric boundary layer.

Assuming a constant OH radical concentration [OH], then equation (1) can be integrated to give:

$$[OC] = [OC]^{\circ}\exp(-k[OH]t) \qquad (25)$$

where $[OC]^{\circ}$ is the initial concentration of the organic compound as the air parcel left the urban or industrial region. Assuming that the OH radicals are efficiently recycled by reaction with the organic compound then we can substitute equation (25) in equation (24) to generate:

$$d/dt[O3] = \alpha k[OH][OC]^{\circ}\exp(-k[OH]t) - [O_3]Vg/h \qquad (26)$$

This equation may be integrated by parts, using the integrating factor method, to yield:

$$[O_3]\exp(Vgt/h) = \alpha k[OH][OC]^{\circ}\exp((k - Vg/h)t)/(k - Vg/h) + [O_3]^{b} \qquad (27)$$

where $[O_3]^{b}$ is the baseline O_3 concentration when the air parcel first entered the urban and industrial region. Rearranging this equation, we derive the following expression for the time development of the O_3 concentration in the air parcel as it moves into the downwind environment:

$$[O_3] = \alpha k[OH][OC]^{\circ}(\exp(-Vgt/h) - \exp(-k[OH]t))/(k - Vg/h) + [O_3]^{b}\exp(-Vgt/h) \qquad (28)$$

The downwind O_3 concentration is thus the sum of two terms, a regional photochemical enhancement term and a transported baseline term. Our interest here is primarily in the regional photochemical enhancement (RPE) term which is given by:

$$RPE = \alpha k[OH][OC]^{\circ}(\exp(-Vgt/h) - \exp(-k[OH]t))/(k[OH] - Vg/h) \qquad (29)$$

The RPE is accordingly dependent on α, the number of O_3 produced per OC oxidised, the rate coefficient k for the reaction of OH with OC, the emission rate of OC through [OC]° and the deposition velocity for O_3. Because it is the difference between two exponential terms, it rises to a maximum and then decays away.

The time of the RPE maximum, t_{max}, depends on the reactivity of the organic compound and the deposition time constant of O_3 but is independent of the number of O_3 molecules produced by the reaction of OH with the organic compound. For our example organic compound ethylene, we can take $k = 8.5 \times 10^{-12}$ $cm^3 molecule^{-1} s^{-1}$, $[OH] = 5 \times 10^6$ molecule cm^{-3}, Vg = 5 mm s^{-1} and h = 1300 m (Smith and Hunt, 1978), gives the maximum of the RPE after 15 hours of travel. After this travel time has elapsed, the air parcel may have travelled 100s of kilometres and may have crossed national or international boundaries in doing so. However, if the organic compound was the highly reactive isoprene from biogenic sources, we would have $k = 1.0 \times 10^{-10}$ cm^3 $molecule^{-1} s^{-1}$ which would give t_{max} = 2.5 hours. In this case, the RPE maximum would still be in the locality of the source region.

The Python code illustrating regional photochemical ozone formation over Europe is available in annexIII_exerciseIII3.pdf.

REFERENCES

Grennfelt, P., Hov, O., Derwent, R., 1994. Second generation abatement strategies for NO_x, NH_3, SO_2 and VOCs. *Ambio* 23, 425–433.

NAEI, 2021. *Air pollutant inventories for England, Scotland, Wales and Northern Ireland: 2005–2019.* Report ED11787. Didcot: Ricardo Energy & Environment.

Metcalfe, S.E., Atkins, D.H.F., Derwent, R.G., 1989. Acid deposition modelling and the interpretation of the United Kingdom secondary precipitation network data. *Atmospheric Environment* 23, 2033–2052.

Metcalfe, S.E., Whyatt, J.D., Derwent, R.G., 1995. A comparison of model and observed network estimates of sulphur deposition across Great Britain for 1990 and its likely source attribution. *Quarterly Journal of the Royal Meteorological Society* 121, 1387–1411.

Smith, F.B., Hunt, R.D., 1978. Meteorological aspects of the transport of pollution over long distances. *Proceedings of the International Symposium Held in Dubrovnik, Yugoslavia, 7–14 September 1977*, pp. 461–499. www.sciencedirect.com/article/pii/B9780080229324500513.

APPENDIX 4: GLOBAL CLIMATE CHANGE MODELS

In this appendix, we describe how to model the global climate change from a range of atmospheric trace gases including carbon dioxide (CO_2), methane (CH_4), tropospheric ozone (O_3), nitrous oxide (N_2O), halocarbons and from aerosols. Amalgamating the model predictions from all these greenhouse gases and aerosols allows us to estimate the rise in global temperature since pre-industrial times and to compare our predictions with the observed global temperature record. As we will show, our predictions are accurately borne out by the observations, giving us confidence in our model approach.

Global climate change is a complex phenomenon so we begin by identifying the key processes involved and come to grips with their timescales. Since the industrial revolution, mankind has changed the composition of the atmosphere through emissions of certain key air pollutants. These pollutants, the greenhouse gases and aerosols have intercepted the outgoing infrared radiation leaving Earth and disturbed Earth's radiation balance with the incoming radiation from the sun. The climate system has responded to this out-of-balance by warming Earth's surface and atmosphere and so regaining the radiation balance. This warming is the so-called greenhouse effect and it is said to have been forced by the atmospheric composition change. This forcing mechanism is called radiative forcing and the Earth–atmosphere system response is climate change.

The key processes involved with the radiative forcing of global climate change are illustrated in the framework diagram shown in Figure A4.1. The arrows pointing downwards represent cause and

EMISSIONS FROM HUMAN ACTIVITIES

⇓

ATMOSPHERIC CONCENTRATION CHANGES

⇓

RADIATIVE FORCING

⇑⇓

PHYSICAL RESPONSE

⇓

CLIMATE RESPONSE

⇓

BIOLOGICAL RESPONSE

Figure A4.1 Framework diagram representing the key processes involved with the radiative forcing of climate change.

effect linkages and illustrate the various forcing mechanisms. Emissions of greenhouse gases and aerosols from human activities cause changes in the composition of the atmosphere. These atmospheric composition changes, since pre-industrial times, interfere with Earth's radiation balance and lead to radiative forcing. Earth's system responds to this radiative forcing by warming and this is the physical response shown in the framework diagram above.

The approach to modelling the global climate system we have used in this monograph is based on the simple conceptual model described by Hartmann (1994). We begin by assuming that the oceans can be represented by a wet surface with the heat capacity of the top or mixed layer of the oceans. We do not represent the ocean currents and their slow response to the radiative forcing and so we will be unable to describe regional variations in the temperature response. Nevertheless, this slab-ocean description is a reasonable starting point. Introducing the greenhouse gases into the atmosphere has led to a radiative forcing of ΔRF in W m^{-2} since pre-industrial times and we would like to estimate the time, t, dependence of the global atmospheric temperature rise ΔT in Kelvin (or degrees Celsius) that has resulted. To do this, we use Hartmann's equation:

$$C d\Delta T/dt = \Delta RF - \Delta T/\lambda \qquad (1)$$

where C is the heat capacity of the ocean and λ is the climate sensitivity. The term on the left-hand side of equation (1) represents the storage of heat energy in the ocean.

We rearrange equation (1) and put it in its standard form:

$$d\Delta T/dt = \Delta RF/C - \Delta T/\lambda C \quad (2)$$

which can be solved for ΔT by integrated by parts as follows to give:

$$\Delta T = \exp(-t/\lambda C).\int_0^t \Delta RF/C \exp(t/\lambda C).dt \quad (3)$$

If the global system is forced by an *instantaneous switch on of a steady forcing* of ΔRF at time t = 0, then the time dependence of the global temperature rise is given by:

$$\Delta T = \lambda \Delta RF\,[1 - \exp(-t/\lambda C)] \quad (4)$$

so that the equilibrium temperature rise $\Delta T = \Delta RF_0\lambda$ is approached exponentially with a time constant of λC. If the heat capacity of the oceans is assumed to be that of 75 metres of ocean, then $C = 3.15 \times 10^8$ J K^{-1} m^2 (Hartmann, 1994). The climate sensitivity may be taken to be $\lambda = 0.526$ K/W m^{-2} (Hartmann, 1994), corresponding to a 2.0°C rise in global atmospheric temperature for a doubling in CO_2. On this basis, the equilibrium temperature rise is approached with a time constant $\lambda C = 1.66 \times 10^8$ J W^{-1} = 1.66×10^8 seconds = 5.3 years, because there are 3.1536×10^7 seconds in a year and a Watt is a Joule s^{-1}. This gives us a measure of the time constant in the relationship between climate forcing and global warming.

If we apply a forcing that increases linearly in time from zero at t = 0 of the form

$$\Delta RF = Qt \quad (5)$$

then substituting in equation (3) and integrating gives

$$\Delta T = \lambda Q\,[t + \lambda C(\exp(-t/\lambda C) - 1)] \quad (6)$$

which for $t >> \lambda C$ gives:

$$\Delta T = \lambda Q\ (t - \lambda C) \qquad (7)$$

Equation (7) means that the time-dependent temperature response ΔT is simply the equilibrium response λQt delayed by λC or six years. In this way, we get a feel for the magnitude of the time delays in the atmosphere-ocean system between forcing and temperature response.

The equilibrium temperature response from doubled carbon dioxide levels is termed the *equilibrium climate sensitivity*. This equilibrium temperature response provides a benchmark against which global climate models can be compared. In our climate model, the radiative forcing from doubled carbon dioxide levels is given by $5.49 \times \ln 2 = 3.8\ Wm^{-2}$ and hence the equilibrium climate sensitivity from equation (4) is found to be $\Delta T = \lambda \Delta RF = 2.0°C$. This value is well within the range of 1.5–4.5°C given by the Intergovernmental Panel on Climate Change (2013) in their chapter on long-term climate change projections, commitments and irreversibility. This range is unchanged between the First Assessment Report (Intergovernmental Panel on Climate Change, 1990) and the Fifth Assessment Report (Intergovernmental Panel on Climate Change, 2013).

For most applications of our climate model, we need *a more general solution* to Hartmann's equation so that we can represent the time history of the greenhouse gas concentrations and hence their radiative forcing. We therefore break up the time history of the radiative forcing into a set of one-year steps and calculate the global temperature rise at the end of every year. Hence, for an annual step in radiative forcing of ΔRF and using equation (3), the global temperature rise, ΔT, builds up to a value of $\Delta T^{t=1} = \lambda \Delta RF\ [1 - \exp(-1/\lambda C)]$ at the end of the first year, before decaying away exponentially. The value of the bracket term is 0.17 and so $\Delta T^{t=1} = 0.090 \Delta RF$. The time constant of the exponential decline is given in equation (1) and is found to be λC, that is to say, about six years.

A time profile of the global temperature rise forced by a trace gas can therefore be constructed by splitting it into a set of annual steps, and using the ΔRF and the exponential decline to represent the temperature rise from each year, then summing over the whole time series, backwards in time. We illustrate this procedure with a number of worked exercises as detailed below.

EXERCISE A4.1 THE CONTRIBUTION FROM CARBON DIOXIDE TO GLOBAL WARMING

In this exercise, we calculate the contribution from carbon dioxide (CO_2) to global warming using a set of annual steps beginning in 1751 and finishing in the present day in the 21st century. We use as a starting point, the time history of the annual global CO_2 emissions from anthropogenic activities from Boden et al., (2017). Annual anthropogenic emissions have grown from 11 million tonnes CO_2 per year in 1751 to 36 billion tonnes per year in 2014. These annual emissions, E, produce increments in the global mean CO_2 level of αE, where $\alpha = 0.128$ ppm per billion tonnes per year, based on a CO_2 molar mass of 44.0 g mol^{-1} (Kaye and Laby, 1966), an atmospheric molar mass of 28.9644 g mol^{-1} (National Oceanic and Atmospheric Administration, 1976) and an atmospheric mass of 5.136×10^{21} g (Trenberth et al., 1987), assuming that the CO_2 is instantaneously mixed throughout the atmosphere.

Once in the atmosphere, the anthropogenic CO_2 exchanges with the biosphere and the top-layer of the oceans and absorbs incoming and outgoing infrared radiation, thus driving the radiative forcing of the global climate system. These exchange processes deplete the atmospheric burden such that, of the anthropogenic CO_2 emitted during one year, a fraction, f, remains airborne after t years. In this exercise, we will represent the airborne fraction f by

$$f = a_1\exp(-t/b_1) + a_2\exp(-t/b_2) + a_3\exp(-t/b_3) \qquad (8)$$

In this exercise we will take $a_1 = 0.30030$, $a_2 = 0.34278$, $a_3 = 0.35686$, $b_1 = 6.993$ years, $b_2 = 71.109$ years and $b_3 = 815.727$ years, based on the uptake of a pulse of CO_2 by the outcrop-diffusion ocean carbon-cycle model of Sieganthaler (1983).

The anthropogenic CO_2 level in a given year can be obtained by summing the values of αEf for each year backwards in time to 1751. The total CO_2, $[CO_2]$, mixing ratio in a given year is then obtained by adding the anthropogenic CO_2 level to the pre-industrial level $[CO_2]^\circ$ of 279 ppm (Meinshausen et al., 2017) and this is shown in Figure A4.2. Also shown in this figure is the observed record of historic CO_2 mixing ratios taken from ice core records from Antarctica taken from Meinshausen et al., (2017). The agreement is satisfactory for the present purposes.

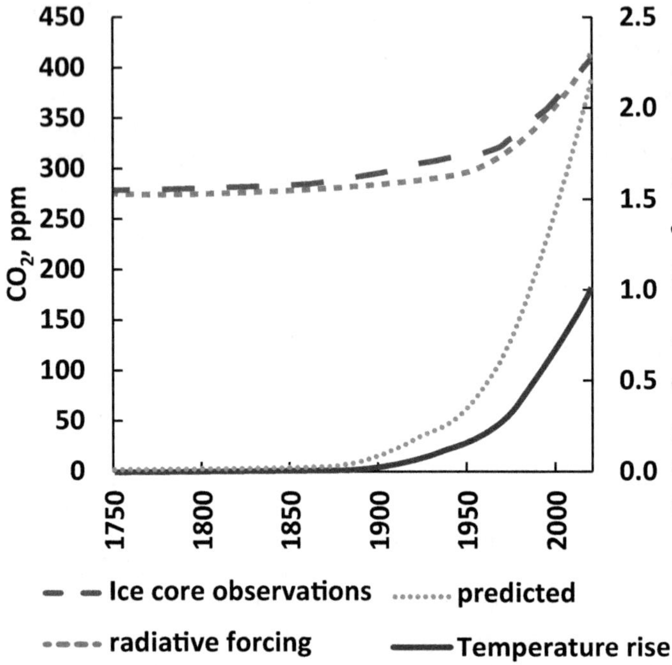

Figure A4.2 Time histories of the global CO_2 mixing ratios in ppm from our predictions and from the ice core reconstructions together with the radiative forcing in W m^{-2} and the rise in global temperature in degrees Celsius.

Radiative forcing is an instantaneous property of the global climate system and can be estimated for each year from 1751 to the present day using $\Delta RF = 5.49 \ln([CO_2]/[CO_2]°)$ in W m^{-2} (Intergovernmental Panel on Climate Change, 2013). The time history of the ΔRF values are also shown in Figure A4.2.

We now have a set of annual ΔRF values that we can feed into the Hartmann equation (1) above to calculate the required time history of the global temperature rise since pre-industrial times, ΔT. We do this by taking a year at a time and calculating the value of ΔT at the end of that year using $\Delta T^{t=1} = \Delta RF°\lambda\ [1 - \exp(-1/\lambda C)] = 0.090\Delta RF°$ from above and calculating the value of $\Delta T^{t} = 0.090\Delta RF°.\ \exp(-t/\lambda C)$, at the end of subsequent years, t years forwards in time, assuming an exponential decay with a time constant of $\lambda C = 5.3$ years. The temperature rise in a given year is then obtained by summing the ΔT^t values for each year backwards in time to 1751. The resulting rise in global temperature since 1751, the required global warming, is shown in Figure A4.2. Global warming from CO_2 reached 1.04°C in 2020.

We see from Figure A4.2 that the rise in atmospheric CO_2 levels is roughly exponential in shape after 1950 or thereabouts and drives a roughly linear increase in radiative forcing because of the logarithmic relationship between radiative forcing and atmospheric mixing ratio. This roughly linear increase in radiative forcing in turn drives a near-linear increase in global temperature. Extrapolating the linear regions of each curve backwards in time to the × axis shows how the global temperature curve lags the rise in atmospheric levels by about 10–20 years.

A computer code in Python (annexiv_exerciseiv1.pdf) is available, providing a detailed description and implementation of this global CO_2 climate model exercise.

EXERCISE A4.2 THE CONTRIBUTION FROM METHANE TO GLOBAL WARMING

In this exercise, we employ an annual step-wise approach, as in the previous exercise, to calculate the contribution from methane (CH_4) to global warming.

CH_4 is a well-mixed trace gas in the troposphere with an atmospheric lifetime of about 12 years due to its destruction by chemical reaction with hydroxyl (OH) radicals (Prather et al., 2012). This reaction depletes the steady state distribution of OH so that the atmospheric lifetime of CH_4 has increased steadily from just under 9 years in pre-industrial times to about 12 years in the present-day. In this way, CH_4 reduces the efficiency of its own removal mechanism. Global mean CH_4 levels have risen from about 730 ppb in pre-industrial times just over 1800 ppb today Meinshausen et al., (2017), driven by anthropogenic CH_4 emissions.

The atmospheric lifetime, τ, fixes the fractions, f, of the anthropogenic CH_4 emitted in each year that remain in the atmosphere in subsequent years, after the time, t, in years, where $f = \exp(-t/\tau)$. The atmospheric lifetime changes each year as the CH_4 builds up in the atmosphere and this is represented using the feedback factor, ff, (Isaksen and Hov, 1987) where ff = fractional change in CH_4 loss rate/fractional change in CH_4 level. Here, we use a feedback factor of –0.29 to account for the change in atmospheric lifetime between 1751 and the present-day.

We use as a starting point, the time history of the annual global CH_4 emissions from anthropogenic activities from Hoesly et al., (2018). Annual anthropogenic emissions have grown from 22 million tonnes CH_4 per year in 1751 to 390 million tonnes per year in 2014 (Hoesly et al., 2018). These annual emissions, E, produce increments in the global mean CH_4 level of αE, where α = 0.352 ppb per million tonnes per year, based on a CH_4 molar mass of 16.04303 g mol (Kaye and Laby, 1966), an atmospheric molar mass of 28.9644 g mol^{-1} (National Oceanic and Atmospheric Administration, 1976) and an atmospheric mass of 5.136×10^{21} g (Trenberth et al., 1987), assuming that the CH_4 is instantaneously mixed throughout the atmosphere.

Anthropogenic activities have driven up atmospheric levels of CH_4. Once in the atmosphere, CH_4 absorbs incoming solar infra-red radiation thus driving the radiative forcing of the global climate system. To estimate the contribution to radiative forcing from CH_4, we must first quantify the growth in the global mean CH_4 levels since pre-industrial times. The anthropogenic CH_4 level in a given year was calculated by summing the fractions, $f\alpha E$, of the anthropogenic emission CH_4 increments remaining in the atmosphere in each year back to 1751. To arrive at the total CH_4 levels, $[CH_4]$, the anthropogenic CH_4 levels are added to the pre-industrial level $[CH_4]^\circ$ and the results are shown in Figure A4.3. Also shown in Figure A4.3 is the observed CH_4 record obtained from the reconstruction of the measurements of CH_4 trapped ice cores from Antarctica (Meinshausen et al., 2017). The agreement is satisfactory for our present purposes.

Radiative forcing is an instantaneous property of the global climate system and can be estimated for each year from 1751 to the present day using $\Delta RF = 0.036(\sqrt{[CH_4]}-\sqrt{[CH_4]^\circ})$ in W m^{-2}. The time history of the ΔRF values are also shown in Figure A4.3.

We now have a set of annual ΔRF values that we can feed into the Hartmann equation (1) above to calculate the required time history of the global temperature rise since pre-industrial times,

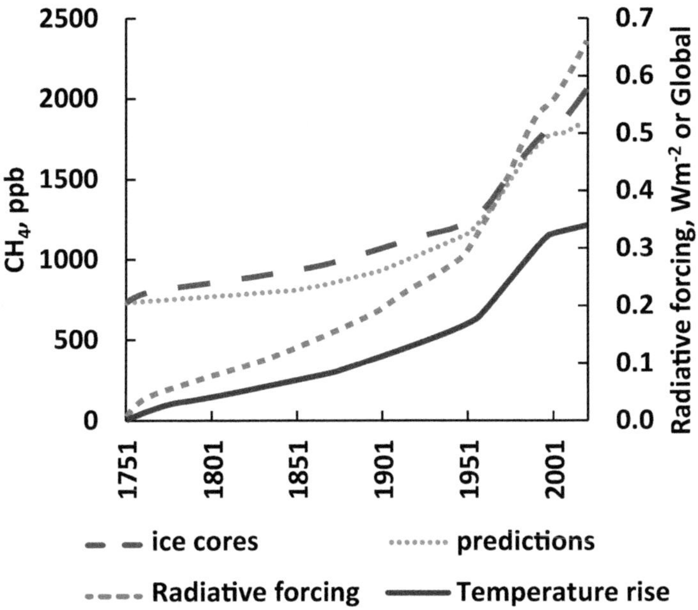

Figure A4.3 Time histories of the global CH_4 mixing ratios in ppb from our predictions and from the ice core reconstructions together with the radiative forcing in W m^{-2} and the rise in global temperature in degrees Celsius.

ΔT, due to CH_4. We do this by taking a year at a time and calculating the value of ΔT at the end of that year using $\Delta T^{t=1} = \Delta RF°\lambda\ [1 - \exp(-1/\lambda C)] = 0.090\Delta RF°$ from above and calculating the value of $\Delta T^t = 0.090\Delta RF°.\ \exp(-t/\lambda C)$, at the end of subsequent years, t years forwards in time, assuming an exponential decay with a time constant of 5.3 years. The temperature rise in a given year is then obtained by summing the ΔT^t values for each year backwards in time to 1751. The resulting rise in global temperature since 1751, the required global warming, is shown in Figure A4.3. Global warming from CH_4 reached 0.34°C in 2020.

A computer code in PYTHON is available (annexIV_exerciseIV2.pdf), providing a detailed description and implementation of this exercise.

EXERCISE A4.3 THE CONTRIBUTION FROM NITROUS OXIDE TO GLOBAL WARMING

In this exercise, we employ the same step-wise approach as in the previous exercises to calculate the contribution from nitrous oxide (N_2O) to global warming. The main difference between this and the previous exercises is that global emissions of N_2O are poorly known and this is particularly the case for the emissions from anthropogenic activities. As a consequence, a radically different approach is required in this exercise.

Nitrous oxide is a long-lived and well-mixed trace gas in the troposphere. From the measurements of N_2O trapped in ice cores from Antarctica it has been possible to reconstruct the time history of atmospheric levels of N_2O over the last thousands of years (Meinshausen et al., 2017). Pre-industrial levels have been accurately characterised at about 273 ppb and the onset of the influence of anthropogenic activities can be traced to the middle of the 19th century. This time series of historic global N_2O levels is the main building block of our approach.

Although the global N_2O emission source strength, E, is not well characterised, its loss processes are well understood. The main loss process for N_2O is stratospheric photolysis and its atmospheric lifetime, τ, has been fixed at 121 years (Intergovernmental Panel on Climate Change, 2013). Setting up an annual global budget for N_2O, we can write for each yearly timestep, Δt:

$$\Delta c = \alpha E \Delta t - c\Delta t/\tau \qquad (9)$$

where $\alpha = 0.1281$ ppb per million tonnes per year, is the conversion factor between atmospheric emission and atmospheric mixing ratio, c, calculated using a molar mass of 44.0128 for N_2O (Kaye and Laby, 1966). Rearranging equation (8), we find:

$$E = (\Delta c/\Delta t + c/\tau)/\alpha \qquad (10)$$

which allows us to convert the observed time history of c into a time history of total emissions. Since we begin our analysis in 1850, the emission calculated for that year is assumed to be the natural emission source strength. The anthropogenic emission source is calculated from the difference between the value calculated for that year using equation (9) and the natural emission source calculated for 1850.

The fraction, f, of the anthropogenic N_2O remaining airborne after t years is given by $f = \exp(-t/\tau)$. Hence, we can calculate the level of anthropogenic N_2O in any year by summing the fractions remaining airborne = αfE of each annual emission backwards in time to 1850. The global mean N_2O level is then obtained by adding the pre-industrial level to the anthropogenic level and these predictions are presented in Figure A4.4. This figure also presents the observations and there is a reasonable correspondence between our predictions and the observed levels from the ice core reconstructions.

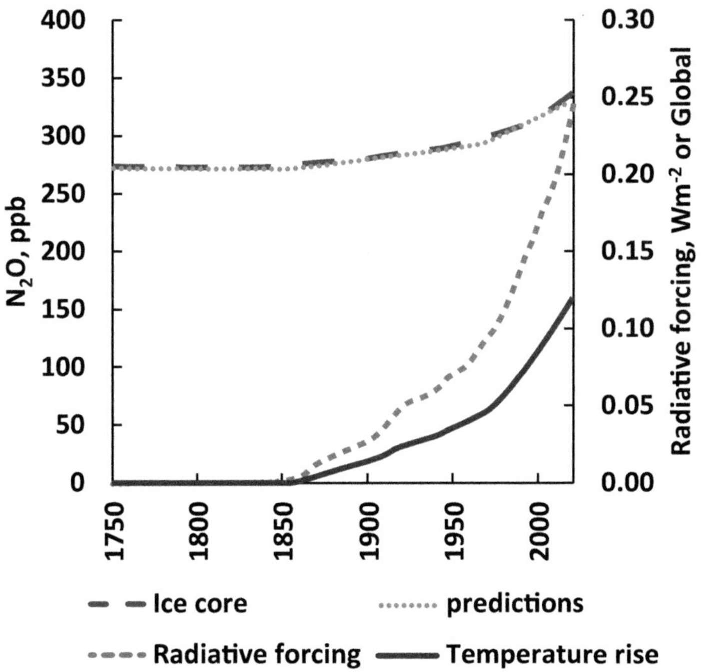

Figure A4.4 Time histories of the global N_2O mixing ratios in ppb from our predictions and from the ice core reconstructions together with the radiative forcing in W m^{-2} and the rise in global temperature in degrees Celsius.

Radiative forcing is an instantaneous property of the global climate system and can be estimated for each year from 1751 to the present day using $\Delta RF = 0.14(\sqrt{[N_2O]} - \sqrt{[N_2O]^\circ})$ in W m^{-2} (Intergovernmental Panel on Climate Change, 2013). The time history of the ΔRF values are also shown in Figure A4.4.

We now have a set of annual ΔRF values that we can feed into the Hartmann equation (1) above to calculate the required time history of the global temperature rise since pre-industrial times, ΔT, due to N_2O. We do this by taking a year at a time and calculating the value of ΔT at the end of that year using $\Delta T^{t=1} = \Delta RF^\circ \lambda\ [1 - \exp(-1/\lambda C)] = 0.090 \Delta RF^\circ$ from above and calculating the value of $\Delta T^t = 0.090 \Delta RF^\circ.\ \exp(-t/\lambda C)$, at the end of subsequent years, t years forwards in time, assuming an exponential decay with a time constant of $1/\lambda C = 5.3$ years. The temperature rise in a given year is then obtained by summing the ΔT^t values for each year backwards in time to 1850. The resulting rise in global temperature since 1850, the required global warming, is shown in Figure A4.4. Global warming from N_2O reached 0.12°C in 2020.

A spreadsheet is available (annexIV_exerciseIV3.xlsx), providing a detailed description and implementation of this exercise.

EXERCISE A4.4 THE CONTRIBUTION FROM TROPOSPHERIC OZONE TO GLOBAL WARMING

In this exercise, we aim to describe the time history of the contribution to global warming from tropospheric ozone. The step-wise approach adopted in Exercises A4.1 to A4.3 does not work with tropospheric ozone. This is because there are no direct emissions of ozone into the atmosphere and all the ozone found there has been formed there by atmospheric chemical reactions. Furthermore, because of its high chemical reactivity, reliable atmospheric measurements have only become available since the 1970s (Tarasick et al., 2019) and there are no ice-core or equivalent records with which to gauge historic levels, as there are with the carbon dioxide, methane and nitrous oxide trace gases considered in Exercises A4.1 to A4.3. As a consequence, a radically different approach has been taken with tropospheric ozone by placing reliance on global ozone models.

Global ozone modelling has a long tradition which spans the growth in our understanding of tropospheric ozone. Initially, box models were used to describe how stratosphere-troposphere exchange and surface depletion by deposition to land surfaces could readily account for observed surface ozone levels. As understanding of photochemical smog formation grew in the 1970s, it became obvious that the atmospheric chemical processes involving CH_4, CO, NO_x and VOCs were active throughout the troposphere and were not restricted to the Los Angeles air basin. Simple box models had to expand into two-dimensional (altitude-latitude) and ultimately into three-dimensional (altitude-latitude-longitude) models to accommodate the growing complexity of tropospheric ozone chemistry.

It is now firmly established that stratosphere-troposphere exchange and surface depletion by deposition to land surfaces are not the most important processes controlling tropospheric ozone sources and sinks, respectively. To these must be added photochemical ozone production and destruction. Whilst these processes are closely in balance and leave the global tropospheric ozone burden largely unchanged, they dramatically increase the tropospheric ozone turnover, hence reducing the ozone atmospheric lifetime from several months to 2–3 weeks (Young et al., 2018).

Three-dimensional models are the main tools for understanding the global distribution of tropospheric ozone. They couple together our understanding of transport processes, atmospheric chemistry and the emissions of the CH_4, CO, NO_x and VOCs which act as ozone precursors. Initially, these models were classed as chemistry-transport models, operating using off-line meteorological databases. More recently, with the availability of increasing computer resources, they have been absorbed into global climate models, producing Earth system models (ESMs). ESMs have a number of advantages over chemistry-transport models. ESMs have the advantage of providing access to on-line meteorological databases within the ESM which offer a seemless approach spanning the entire period from pre-industrial times in the 1750s through to 2100. They have the additional advantage of providing improved representations of stratosphere-troposphere exchange and of the emissions of natural biogenic organic compounds through access to the coupled climate model. It has to be said, however, that ESMs are fearfully complex and are only as good as the individual process representations with which they are constructed.

Global ozone models, whether chemistry-transport models or ESMs, predict a doubling of tropospheric ozone levels since pre-industrial times. Unfortunately, there are no observations to confirm or deny these predictions (Tarasick et al., 2019). The consequence of this doubling is a significant contribution to the radiative forcing of the global climate system which makes tropospheric ozone the third most important man-made greenhouse gas after carbon dioxide and methane. Tropospheric ozone has a high radiative efficiency in terms of radiative forcing per ppb increase because of its strong absorptions in the ultra-violet, visible and infra-red regions of the spectrum.

One of the strongest drivers of the increase in global tropospheric ozone is the growth in historic man-made NO_x emissions. In this exercise, we use the Intergovernmental Panel on Climate Change (2013) literature estimate of the 2011 radiative forcing for tropospheric ozone of 0.40 ± 0.2 W m^{-2} as a benchmark and scale to other years based on the global man-made NO_x emissions for that year. Global man-made NO_x emissions are available from Hoesly et al., (2018).

The resulting time history of tropospheric ozone radiative forcing is shown in Figure A4.5. A time profile of the global temperature rise forced by tropospheric ozone was constructed by splitting it into a set

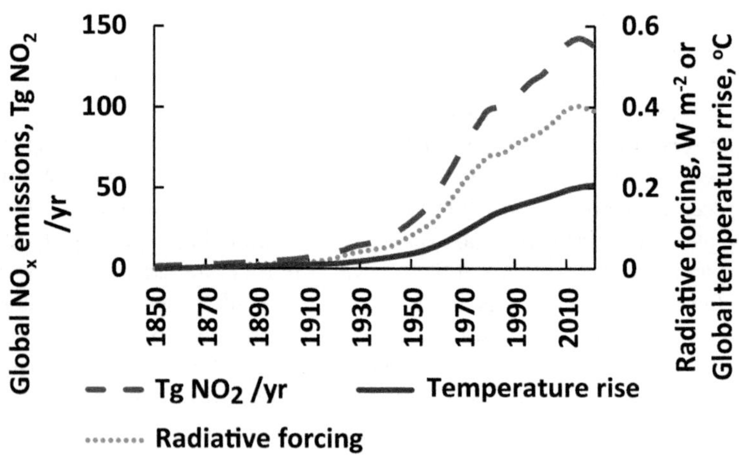

Figure A4.5 Time histories of the global man-made NO_x emissions, the radiative forcing from tropospheric ozone and the rise in global temperature that it drives from 1850 to 2020.

of annual steps, ΔRF. Using ΔRF and the general solution of Hartman's equation from above to represent the temperature rise from each year, we then sum over the whole time series, backwards in time. The resulting time history of the rise in global temperature produced by the radiative forcing from tropospheric ozone is also shown in Figure A4.5. Global warming from tropospheric ozone reached 0.21°C in 2020.

Details of these calculations are provided in a spreadsheet (annexIV_exerciseIV4.xlsx) to which the reader is referred for further information.

EXERCISE A4.5 THE CONTRIBUTION FROM HALOCARBONS TO GLOBAL WARMING

In this exercise, we calculate the contribution to global warming from 35 selected halocarbons. Emissions data for this wide range of halocarbons are generally either unreliable or unavailable so we have a different approach to the other exercises based on the observed global mean halocarbon concentrations. All of these halocarbons are entirely novel, man-made compounds that have only recently been emitted into the atmosphere and so have been present in the atmosphere only from the 1940s onwards (Meinshausen et al., 2017). The only exception is carbon tetrafluoride (CF_4), a perfluorocarbon which is emitted historically by volcanic action, in addition to its current man-made source from the electrolysis process in aluminium manufacture. Our methodology is able to deal with CF_4 since there is no evidence of recent, post-1940, volcanic sources of CF_4 and all the growth in the global mean CF_4 concentration can be attributed to human activities. Details of the historic global mean halocarbon concentrations for the 35 halocarbons are taken from Meinshausen et al., (2017).

For each of the 35 halocarbons, the time history of its radiative forcing was calculated by scaling the historical global mean concentration in ppb with its radiative efficiency defined as the radiative forcing per ppb taken from Intergovernmental Panel on Climate Change (2013). The individual radiative forcings were summed to give the total radiative forcing for all 35 halocarbons. The total radiative forcing caused a rise in global temperature which was estimated using the general solution to the Hartman equation as outlined above. The rise in global temperature from all 35 halocarbons is plotted in Figure A4.6, together with the contributions from 5 selected halocarbons that made substantial contributions during the period from 1940–2020.

Halocarbons, as a class, contribute about 0.2°C to global warming in 2020 as shown in Figure A4.6. The two most important halocarbons currently are CFC-12 (CF_2Cl_2, difluorodichloromethane) and CFC-11 (CCl_3F, trichlorofluoromethane), which were heavily used as refrigerants and aerosol propellants, respectively. Their manufacture and use have been phased out under the terms of the Montreal Protocol and its Subsequent Amendments, but, because of their long atmospheric lifetimes, they will continue to contribute strongly to global warming for decades to come (World Meteorological Organization, 2014). The third most important halocarbon is HCFC-22 (CHF_2Cl, chlorodifluoromethane) a widely air conditioning fluid for application in buildings. Its use is still growing, hence also its contribution to global

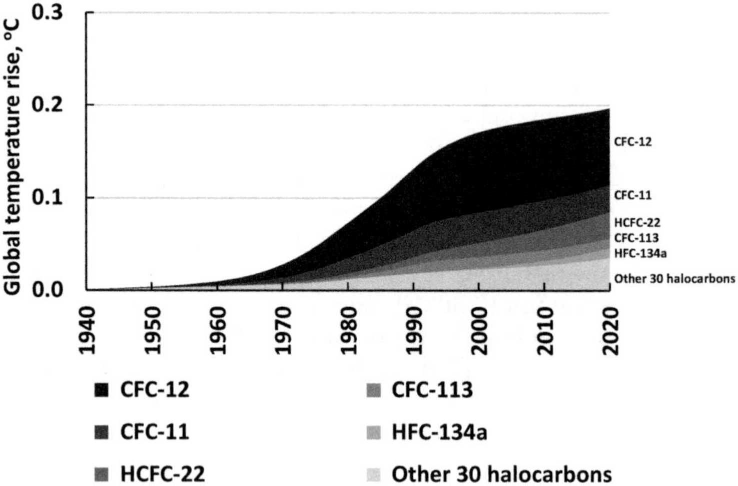

Figure A4.6 Time histories of the contributions to the rise in global temperature caused by 35 halocarbons and attributed to the five most important halocarbons.

warming as shown in Figure A4.6, despite its phase-out in developing countries under the Montreal Protocol. Fourth in the rankings of the halocarbon global warming is CFC113 (CCl_2FCF_2Cl, 1,1,2-trichloro1,2,2-trifluoroethane), a widely used solvent for cleaning electrical components and electronic circuit boards, which also has been phased out under the terms of the Montreal Protocol. In fifth place is HFC–134a (CF_3CHF_2, 1,1,1,2-tetrafluoroethane), a widely used mobile air conditioning fluid and refrigerant, with the latter use as a replacement for CFC12. However, because of its contribution to global warming, it is also being phased out and is being replaced by the hydrofluoroolefin HFO–1234yf ($CH_2 = CFCF_3$, 2,3,3,3-tetrafluoropropene) in mobile air conditioning applications. Together, these top-five halocarbons account for 0.16°C global warming, leaving the remining thirty to account for the remaining 0.035°C.

Details of these model calculations are provided in a spreadsheet (annexIV_exerciseIV5.xlsx) to which the reader is referred for further information.

EXERCISE A4.6 THE CONTRIBUTION FROM AEROSOLS TO GLOBAL WARMING

In this exercise, we estimate the contribution to global warming from aerosol particles.

The impact of aerosol particles on the global climate system offers several contrasts to that of the well-mixed greenhouse gases considered in Exercises A4.1 to A4.3 above. Aerosol particles have lifetimes of 1–2 weeks compared with 10s to 100s of years for the greenhouse gases and so exhibit highly variable concentrations in the lower atmosphere. Aerosol particles largely act to cool the global climate system in contrast to the greenhouse gases because the former scatter incoming radiation (Intergovernmental Panel on Climate Change, 1990). This cooling effect turns out to have crucial importance because it partially offsets the global temperature rise from the greenhouse gases. Without this offset, we would greatly overpredict the present-day global temperature rise and we would not be able to explain the shape of the time history of its development. It is therefore crucial to have an accurate representation of the cooling from aerosol particles in global climate models.

It has long been recognised that a large proportion of the aerosol particles that influence the radiation balance in the atmosphere and affect cloud processes are derived from man-made sulphur emissions. Man-made SO_2 emissions have increased dramatically in recent years, causing increases in aerosol sulphate loadings over the industrialised regions of North America and Europe and contributing to regional acidification problems, see Chapter 5. From analyses of Greenland ice cores, it has been shown that levels of sulphate and nitrate have been increasing since pre-industrial times. However, there are no long-term continuous observations of aerosol parameters globally with which to construct a time history of aerosol radiative forcing (Intergovernmental Panel on Climate Change, 1990).

In this exercise, we use historic global SO_2 and NO_x emissions taken from Hoesly et al., (2018) to guide the shape of the time history of the radiative forcing from aerosol particles and scale it to a literature estimate for 2010 of –0.9 W m^{-2} taken from Intergovernmental Panel on Climate Change (2013). The aerosol radiative forcing causes a decrease in global temperature which was estimated using the general solution to the Hartman equation as outlined above. The time history of the radiative forcing and decrease in global temperature are plotted out in Figure A4.7.

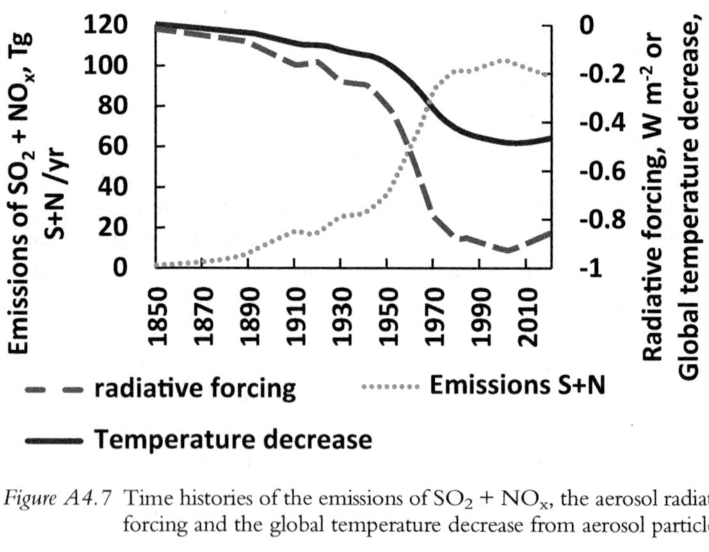

Figure A4.7 Time histories of the emissions of $SO_2 + NO_x$, the aerosol radiative forcing and the global temperature decrease from aerosol particles.

The time history of the global cooling from aerosol particles in Figure A4.7 builds up during the twentieth century and reaches −0.46°C in 2000. There is evidence that, following the downturn in global SO_2 and NO_x emissions as coal use is phased out, global cooling diminishes in the 2010s.

Details of these calculations are provided in a spreadsheet (annexIV_exerciseIV6.xlsx) to which the reader is referred for further information.

We cannot, however, pretend that Figure A4.7 is anything other than a minimal treatment of the real impact of aerosol particles on global climate change. There are other aerosol components in addition to aerosol sulphate and nitrate and other climate-aerosol interactions in addition to the scattering and absorption of radiation.

Black carbon (BC) is an example of an aerosol component that darkens snow and ice surfaces and makes them less reflective, leading to increased melting and higher polar temperatures. These impacts are dramatic in the Arctic but have had a relatively weak impact on the global temperature rise since pre-industrial times.

Aerosol-cloud impacts involve changes in cloud droplet size distributions, cloud amounts, cloud brightness and cloud lifetimes driven by the impact of man-made aerosol particles. Because of the lack of global and continuous observations of aerosols and their properties, it has been difficult to quantify the radiative forcing from aerosol-cloud interactions. Estimates from the IPCC put the radiative forcing from aerosol-cloud interactions on a par with those from aerosol-radiation, though both have large uncertainty ranges.

In this exercise, we have used a benchmark radiative forcing of -0.9 W m^{-2} in 2010 which includes -0.45 W m^{-2} from each of the aerosol-radiation and aerosol cloud interactions, based on the Intergovernmental Panel on Climate Change (2013) assessments. For further information on the global climate impacts of aerosols, the reader is referred to the outstanding series of IPCC reports.

REFERENCES

Boden, T.A., Marland, G., Andres, R.J., 2017. *Global, regional and national fossil-fuel CO_2 emissions*. Oak Ridge, TN: Carbon Dioxide Information and Analysis Center, Oak Ridge National Laboratory. (doi:10.3334/CDIAC/00001_V2017)

Hartmann, D., 1994. Global physical climatology. In R. Dmowska and J.R. Holton (eds), *International geophysics series volume 56*. San Diego, CA: Academic Press.

Hoesly, R.M., *et al.*, 2018. Historical (1750–2014) anthropogenic emissions of reactive gases and aerosols from the Community Emissions data System (CEDS). *Geoscientific Model Development* 11, 369–408.

Isaksen, I.S.A., Hov, O., 1987. Calculation of trends in the tropospheric concentrations of O_3, OH, CH_4 and NO_x. *Tellus* 39B, 271–285.

Intergovernmental Panel on Climate Change, 1990. *Climate change: The IPCC scientific assessment*. Report prepared for Intergovernmental Panel on Climate Change by Working Group 1. Cambridge: Cambridge University Press.

Intergovernmental Panel on Climate Change, 2013. *Climate change 2013: The physical science basis*. Contribution of Working Group I to the Fifth Assessment Report of the Intergovernmental Panel on Climate Change. Cambridge: Cambridge University Press.

Kaye, G.W.C.K., Laby, T.H., 1966. *Tables of physical and chemical constants*, p. 150. London: Longmans.

Meinshausen, M., *et al.*, 2017. Historical greenhouse gas concentrations for climate modelling (CMIP6). *Geoscientific Model Development* 10, 2057–2116.

National Oceanic and Atmospheric Administration, 1976. *US standard atmosphere, 1976*. Report NOAA-S/T 76-1562. Washington, DC: National Oceanic and Atmospheric Administration.

Prather, M.J., Holmes, C.D., Hsu, J., 2012. Reactive greenhouse gas scenarios: Systematic exploration of uncertainties and the role of atmospheric chemistry. *Geophysical Research Letters* 39, L09803. (doi:10.1029/2012/GL051440)

Sieganthaler, U., 1983. Uptake of excess CO2 by an outcrop-diffusion model of the ocean. *Journal of Geophysical Research* 88, 3599–3608.

Tarasick, D., *et al.*, 2019. Tropospheric ozone assessment report: Tropospheric ozone from 1877 to 2016, observed levels, trends and uncertainties. *Elementa Science of the Anthropocene* 7, 39–110. (doi:10.1525/elementa.376)

Trenberth, K.E., Christy, J.R., Olson, J.G., 1987. Global atmospheric mass, surface pressure and water vapour variations. *Journal of Geophysical Research* 92, 14, 815–814, 826.

World Meteorological Organization, 2014. *Scientific assessment of ozone depletion, 2014*. Global Ozone Research and Monitoring Project report no. 55. Geneva, Switzerland: World Meteorological Organization.

Young, P.J., *et al.*, 2018. Tropospheric ozone assessment report: Assessment of global-scale model performance for global and regional ozone distributions, variability, and trends. *Elementa Science of the Anthropocene* 6, 10–59. (doi:10.1525/elementa.265)

LIST OF PYTHON CODES AND SPREADSHEETS

These Python codes and spreadsheets are available to download from the following web address: http://www.routledge.com/9781032275185

annexI_exerciseI1.pdf Decorating an indoor space

annexI_exerciseI2.pdf A pollutant with both indoor and outdoor sources – Albert Hall

annexI_exerciseI3.pdf Indoor PM levels close to a busy road

annexII_exerciseII1.pdf Airborne concentrations downwind of a single point source of pollution

annexII_exerciseII2.pdf Air pollution concentrations generated by a busy road

annexII_exerciseII3.pdf Leakage of natural gas in the London area

annexIII_exerciseIII1.pdf Sulphur budget for the United Kingdom

annexIII_exerciseII2.pdf Oxidised nitrogen budget for the United Kingdom

annexIII_exerciseIII3.pdf Regional photochemical ozone formation across Europe

annexIV_exerciseIV1.pdf Global CO_2 climate model

annexIV_exerciseIV2.pdf Global CH_4 climate model

annexIV_exerciseIV3.xlsx Global N_2O climate model

annexIV_exerciseIV4.xlsx Global tropospheric O_3 climate model

annexIV_exerciseIV5.xlsx Contributions from halocarbons to global warming

annexIV_exerciseIV6.xlsx Global aerosol-climate model

GLOSSARY

acidification	Acid formation caused by sulphur dioxide and NO_x pollution usually in connection with soil or water.
acid rain	Rainfall combining with sulphur dioxide to give acidic rain.
aerosols	A dispersion of solid or liquid particles in a gas.
air pollution	A combination of chemical substances in the atmosphere which can impair health.
air quality	Describes how polluted the air is we breathe and whether it is clean enough for humans and plants to remain healthy.
air quality model	Tool used by policy-makers to predict the emission changes required to meet a given air quality target.
air quality monitoring	Routine measurement of air pollutants in order to establish whether air quality is improving.
air quality standard	A legal maximum level for the concentration of an air pollutant.
air quality target	Level of air quality that must be achieved in a city, state, region or country.
anthropogenic	Resulting from human activities.
atmospheric boundary layer	Lowest part of the atmosphere from ground level to about one kilometre in height, containing the air that we breathe.

background ozone (O_3)	Ozone arising from natural as distinct from man-made sources.
benzene (C_6H_6)	An aromatic liquid at ambient temperature which readily evaporates and is carcinogenic.
best available technology (BAT)	The most efficient technical means of reducing pollutants or improving plant efficiency, regardless of costs.
biogenic	Natural emissions from trees and plants.
1,3-butadiene	A gaseous hydrocarbon emitted by gasoline vehicles and linked to cancer formation.
carbon canister	A container of activated carbon which absorbs hydrocarbon emissions from gasoline vehicle evaporative losses from fuel tanks and from refilling tanks on petrol station forecourts.
carbon dioxide (CO_2)	The gas formed by the combustion of fuels, oil, coal and gas. The most important man-made greenhouse gas.
carbon monoxide (CO)	A toxic gas formed in combustion chambers due to an insufficient supply of oxygen.
carcinogen	A substance that can cause cancer.
car parc	number and distribution of vehicles in any country with respect to engine size and fuel used.
catalytic converter	device attached to the exhaust system of road vehicles containing a catalyst system for reducing air pollutants.
clean air act	The statute in the United States, published in 1963, governing air quality.
control factor	degree of improvement in any pollution control device.
Convention on Long Range Transboundary Air Pollution (CLRTAP)	United Nations declaration whereby states have an obligation to ensure that activities carried out in one country do not environmentally damage others.

Council of Ministers	Prime ministers or heads of state of the countries of the European Union.
dispersion	process of dispersing air pollutants.
downwind	direction towards which the wind is blowing.
dry deposition	Type of deposition whereby the pollutant is removed from the atmosphere through processes such as impaction, absorption and sedimentation.
ecosystem	A system involving the interactions between a community and its non-living environment.
EC ozone directive	The European Commission Directive calling for a long-term ozone limit of 60 ppb on an 8-hour basis.
EMEP model	Emissions Monitoring and Evaluation Programme air quality model usually associated with ozone, acidification and particle formation.
emission inventory	Information on the emissions from all sources of pollution within a city, region or country, as used in an air quality model.
emissions limit	Usually relates to new vehicles whose engines are manufactured to give measure emissions on or below a limit value.
engine exhaust emissions	Gases emitted from the tailpipes of vehicle engines.
Environmental Protection Agency (EPA)	The United States legislative body set up in 1970 to identify pollutants that may endanger public health and issue appropriate air quality criteria for each pollutant.
Eulerian grid air quality model	A tool used to calculate pollutant concentrations over a fixed region of space using a three-dimensional grid mesh.
European Environment Agency	European body set up in 1990 to collect, process and analyse environmental data, including the development of European emissions inventories.

European Union (EU)	The 27 European Member States.
eutrophication	Over-supply of nutrients to soils leading to the loss of biological diversity and the disappearance of sensitive plants.
global warming	Gradual increase in the temperature of Earth due to the greenhouse effect.
greenhouse gases	Gases that contribute to warming Earth by trapping solar energy. The most important are carbon dioxide, methane, and ozone.
grid	A matrix that divides an area into rectangles or a volume into boxes for calculation or measurement of air pollution.
hazardous air pollutants (HAPs)	A list of toxic air pollutants controlled by the US EPA.
hybrid model	A Eulerian air quality model containing a Lagrangian component.
hybrid vehicle	A combined electric and liquid fuel-powered vehicle. The electric power can be used in urban regions to minimise pollution under low speed driving conditions.
hydrocarbon	Organic compound containing carbon and hydrogen.
integrated assessment model	A policy model which combines the multiple effects of the pollutants NOx, VOCs and sulphur dioxide and the relative cost estimates of different abatement strategies to formulate policies for the improvement of air quality at least cost.
isopleths	Lines of constant ozone level; often the graphical output of an ozone air quality model.
isoprene	One of the most highly reactive ozone precursor gases emitted by trees and also found in gasoline vehicle exhaust.
Lagrangian air quality model	A tool used for the calculation of air pollutant concentrations using air parcels moving with the three-dimensional wind field.

lumped chemistry (reduced air chemistry)	A simplified calculation method whereby VOCs are grouped into chemical classes and a generalised species is used to represent the chemistry of each class. Alternatively, VOCs can be grouped according to their structure or reactivity.
member states	countries or States that form the European Union.
meteorology	Study of Earth's atmosphere, usually related to weather forecasting.
methane	A colourless, odourless and flammable gas which is the second most important man-made greenhouse gas.
mobile sources	All road vehicles.
National Ambient Air Quality Standards (NAAQS)	Legal maximum concentrations of air pollutants as proposed by the US EPA.
National Emissions Ceilings Directive (NECD)	The European Commission Directive for individual reductions of nitrogen oxides, VOCs, and sulphur dioxide from all Member States from 2010.
natural gas	Essentially methane gas which occurs naturally and used as a fuel for both home heating and in power stations for generating electricity.
nitric oxide (NO)	An invisible gas emitted by all combustion systems which rapidly converts to nitrogen dioxide and is not considered harmful to human beings at moderate ambient concentrations.
nitrogen dioxide (NO_2)	A reddish brown gas which can irritate the lung passages and is formed in the atmosphere from the emissions of nitric oxide (NO).
nitrogen oxides (NO_x)	Refers to the mixture of both nitric oxide and nitrogen dioxide which always occur together in the atmosphere.

nitrous oxide (N_2O)	'Laughing gas', a sweet, smelling gas which is used as an anaesthetic in surgery and is the fourth most important man-made greenhouse gas after CO_2, CH_4 and tropospheric O_3.
Non-exhaust particle emissions	Emissions from road vehicles from brake, tyre and road surface wear.
NO_x-limited	describes a region of the atmosphere where reducing NO_x has a greater effect on reducing ground-level O_3 than reducing VOCs.
oxygenates	Organic compounds containing carbon, hydrogen and oxygen, such as alcohols, aldehydes and ketones.
ozone (O_3)	Ground level ozone is formed by sunlight-driven chemical reactions involving hydrocarbons and nitrogen oxide precursors.
ozone layer	Ozone in the stratosphere which is an essential protective layer filtering out the harmful ultraviolet radiation from the sun.
PAN	Peroxyacetylnitrate is a highly reactive organic nitrogen compound formed by the photochemical reactions that drive Los Angeles smog. It is a strong oxidising agent and causes eye-irritation.
particles	Airborne particulate matter.
particulate matter	Small solid substances dispersed in the atmosphere, typically less than 10 microns in diameter, PM_{10} or less than 2.5 microns in diameter, $PM_{2.5}$.
particulate trap	A device fitted into the exhaust system of a diesel engine in order to filter out the particles and so reduce their emission into the atmosphere.
Photochemical Ozone Creation Potential (POCP)	A reactivity index that describes the ability of an organic compound to act as an ozone precursor expressed relative to ethylene.

photochemical reaction	A chemical reaction taking place in sunlight.
photolysis	A chemical decomposition driven by sunlight.
point source	Relates to emissions from a single source fixed in space, such as a refinery or a power station.
polyaromatic hydrocarbons (PAHs)	Organic compounds that are both gaseous and adsorbed onto particulate matter and are composed chemically of two or more benzene rings fused together.
precursor emissions	Volatile organic compounds (VOCs) and nitrogen oxides (NO_x) which combine photochemically in sunshine to form ground-level ozone.
primary particles	Particles emitted directly into the atmosphere, usually from combustion driven processes or wind-driven dust or sea-spray.
RAINS integrated assessment model	A model for EU Member States which combines the multiple effects of the pollutants NO_x, VOCs and sulphur dioxide and the relative cost estimates of different abatement strategies to formulate policies for the improvement of air quality at least cost.
reactive air quality model	A tool combining a meteorology model with an atmospheric chemistry model with the aim of predicting the chemical formation and concentrations of air pollutants.
secondary inorganic aerosols (SIAs)	Particles formed by chemical reactions in the atmosphere which are inorganic in chemical composition, such as ammonium sulphate and ammonium nitrate.
secondary organic aerosols (SOAs)	Particles formed in the atmosphere by chemical reactions involving hydrocarbons and nitrogen oxides in the presence of sunlight which are organic in composition.

secondary particles	Fine particles of 2.5 microns diameter or less formed by sunlight-driven photochemical reactions of other precursor pollutants in the atmosphere.
smog	A combination of smoke and fog.
solvent	Volatile organic compounds used in the home and in industry for decorative paints, vehicle refinishing, dry cleaning and many other applications.
state implementation plans (SIPs)	An individual state's strategy or plan for reducing pollutant levels. Each state in the United States is responsible for setting one.
stationary sources	Relates to emissions sources that are fixed in space. Examples are refineries, power stations and chemical plants.
stratosphere	Layer of the atmosphere that extends from 10 to 50 kilometres above the ground.
stratospheric ozone	Ozone present in the upper atmosphere that forms a protective shield from harmful solar ultraviolet radiation.
sulphur dioxide (SO_2)	A colourless, acidic gas produced by the burning of fossil fuels.
super emitting vehicles	Road vehicles that produce an enormous amount of emissions. They are not always old but often are modern vehicles that have not been well- maintained, or vehicles where the emissions control systems have failed.
three-way catalyst technology	Technology designed to convert exhaust emissions from engine combustion into less damaging pollutants.
titration effect	Atmospheric processes whereby nitric oxide is readily converted to nitrogen dioxide by combining with ozone.
topography	Detailed description of the surface features of a region.

toxic organic micropollutants (TOMPs)	Pollutants emitted by the incomplete combustion of fuels and generally linked to cancer formation.
trajectory	Imaginary three-dimensional path taken by an air parcel containing air pollutants in an air quality model.
trans-boundary air pollution	Air pollutants know no international boundary as they are driven by wind fields from one region or state or country to another.
troposphere	Region bound by the Earth's surface and the base of the stratosphere.
upwind	Towards where the wind is blowing.
vehicle emissions standards	Level of emissions that new vehicles must meet.
volatile organic compounds (VOCs)	A wide range of compounds including hydrocarbons and oxygenates that form ozone in sunlight-driven photochemical reactions in the presence of nitrogen oxides.
wet deposition	Removal of water-soluble pollutants by rain.
WHO guidelines	World Health Organization estimates of maximum concentration levels of an air pollutant which will not impair human health nor damage vegetation.

INDEX

For Product Safety Concerns and Information please contact our EU representative GPSR@taylorandfrancis.com Taylor & Francis Verlag GmbH, Kaufingerstraße 24, 80331 München, Germany

Batch number: 10725833

Printed by Printforce, the Netherlands